Advanced Materials for Demanding Applications

IOP Conference Series: Materials Science and Engineering Volume 74

St. Asaph, United Kingdom
7 – 9 April 2014

Editor:

A. McMillan

ISBN: 978-1-5108-0080-9
ISSN: 1757-8981

Printed from e-media with permission by:

Curran Associates, Inc.
57 Morehouse Lane
Red Hook, NY 12571

Some format issues inherent in the e-media version may also appear in this print version.

Printed by Curran Associates, Inc. (2015)

For permission requests, please contact the Institute of Physics
at the address below.

Institute of Physics
Dirac House, Temple Back
Bristol BS1 6BE UK

Phone: 44 1 17 929 7481
Fax: 44 1 17 920 0979

techtracking@iop.org

Additional copies of this publication are available from:

Curran Associates, Inc.
57 Morehouse Lane
Red Hook, NY 12571 USA
Phone: 845-758-0400
Fax: 845-758-2634
Email: curran@proceedings.com
Web: www.proceedings.com

TABLE OF CONTENTS

IOP Conf. Series: Materials Science and Engineering **74** (2015) 011001 doi:10.1088/1757-899X/74/1/011001

Advanced Materials for Demanding Applications

Co-chairs: Alison McMillan[1] and Stephen Schofield[2]

1 Glyndwr University, UK
2 University College London, UK

Preface. This was a special conference. It was small enough (60+ delegates) but covering a wide range of topics, under a broad end-use focussed heading. Most conferences today either have hundreds or thousands of delegates or are small and very focussed. The topics ranged over composite materials, the testing of durability aspects of materials, and an eclectic set of papers on radar screening using weak ionized plasmas, composites for microvascular applications, composites in space rockets, and materials for spallation neutron sources etc. There were several papers of new characterisation techniques and, very importantly, several papers that started with the end-user requirements leading back into materials selection.

In my own area, there were three talks about the technology for the ultra-precise positioning of individual atoms, donors, and complete monolayers to take modern electronics and optoelectronics ideas closer to the market place.

The President of the Institute opened with an experience-based talk on translating innovative technology into business.

Everyone gave a generous introduction to bring all-comers up to speed with the burning contemporary issues. Indeed, I wish that a larger cohort of first-year engineering PhD students were present to see the full gamut of what takes a physics idea to a success in the market place.

I would urge groups to learn from Prof Alison McMillan (a Vice President of the Institute of Physics) and Steven Schofield, to set up conferences of similar scale and breadth. I took in more than I do from mega-meetings, and in greater depth.

Professor Michael Kelly
Department of Engineering
University of Cambridge.

IOP Conf. Series: Materials Science and Engineering **74** (2015) 011001 doi:10.1088/1757-899X/74/1/011001

Proceedings

As the participants of the conference were from diverse backgrounds, there were different expectations as to the desirability of publishing papers as a Proceedings of this conference. We took the decision to invite participant papers, but not to make it obligatory. In the event, all the papers received were from the "Applications" and "Engineering" end of the spectrum, and it was decided that while the Conference itself was co-chaired, I would be the sole Editor of the Proceedings.

I would hope that in future years, those from the "Materials" and "Physics" end of the spectrum would see greater value in publishing this way, but time will tell. The main objective of any conference is to bring people together, to discuss and share ideas, and to create an environment for the development of working relationships that can lead to ground breaking collaborative research.

I would like to thank all the conference participants and the members of the International Scientific and Local Organising Committees. I would particularly like to thank Claire Garland, Institute of Physics, and Sonja Hardy, Glyndwr University for their support with the conference logistics.

Finally, each paper was peer reviewed by at least two independent subject specialists, so I would like to thank each of the peer reviewers for giving their time, and for their numerous suggestions to authors.

Professor Alison McMillan
Division of Applied Science, Computing and Engineering
Glyndwr University.

Editorial Team

Peer review statement

All papers published in this volume of *IOP Conference Series: Materials Science and Engineering* have been peer reviewed through processes administered by the proceedings Editors. Reviews were conducted by expert referees to the professional and scientific standards expected of a proceedings journal published by IOP Publishing.

IOP Conf. Series: Materials Science and Engineering **74** (2015) 012001 doi:10.1088/1757-899X/74/1/012001

Material issues relating to high power spallation neutron sources

M Futakawa

Materials and Life Science Division, J-PARC Center, Japan Atomic Energy Agency, Tokai-mura, Naka-gun, Ibaraki, 319-1195, Japan.

futakawa.masatoshi@jaea.go.jp

Abstract. Innovative researches using neutrons are being performed at the Materials and Life Science Experimental Facility (MLF) at the Japan Proton Accelerator Research Complex (J-PARC), in which a mercury target system is installed for MW-class pulse spallation neutron sources. In order to produce neutrons by the spallation reaction, proton beams are injected into the mercury target. At the moment, when the intense proton beam hits the target, pressure waves are generated in mercury because of the abrupt heat deposition. The pressure waves interact with the target vessel, leading to negative pressure that may cause cavitation along the vessel wall, i.e. on the interface between liquid and solid metals. On the other hand, the structural materials are subjected to irradiation damage due to protons and neutrons, very high cycle fatigue damages and so-called "liquid metal embrittlement". That is, the structural materials must be said to be exposed to the extremely severe environments. In the paper, research and development relating to the material issues in the high power spallation neutron sources that has been performed so far at J-PARC is summarized.

1. Introduction

Neutrons are used for the innovative research that brings about breakthroughs in scientific and engineering fields, *i.e.* fuel cell, hydrogen embrittlement, protein structure, medicine, *etc*.. High power pulsed-spallation neutron sources are being operated at the "Materials and Life Science Experimental Facility" (MLF) of the "Japan Proton Accelerator Research Complex" (J-PARC) in Japan, and at the "Spallation Neutron Source" (SNS) in the US, which are standing on the way to increase the power up to MW-class [1, 2]. In both facilities, liquid mercury was selected as a target material to produce pulsed-neutrons by spallation reaction because of its high-neutron yield and efficient self-cooling performance, as shown in Figure 1.

At the moment that the protons bombard the mercury, impulsive pressure waves are generated in mercury due to the thermal shock. This results in the process of the pressure wave propagation, which leads to the aggressive cavitation in the mercury that imposes damage on the solid wall of the target vessel [3-7]. The impulsive pressure waves lead to the negative pressure through the propagation in mercury that may cause cavitation. Localized impacts by micro-jets and/or shock waves that are caused by cavitation bubble collapse impose so-called "pitting damage" on the vessel wall. The cavitation phenomenon becomes a crucial obstacle to increasing the power in the mercury target for the pulsed spallation neutron source. In fact, many penetrating pits due to cavitation were found in the inner wall in the used target vessels in the SNS, even under relatively low power [8]. In addition, the structural materials of the target vessels are suffering from damage arising from high intensive neutron and proton irradiation, very high cycle fatigue, liquid metal embrittlement (LME), etc. [9-11].

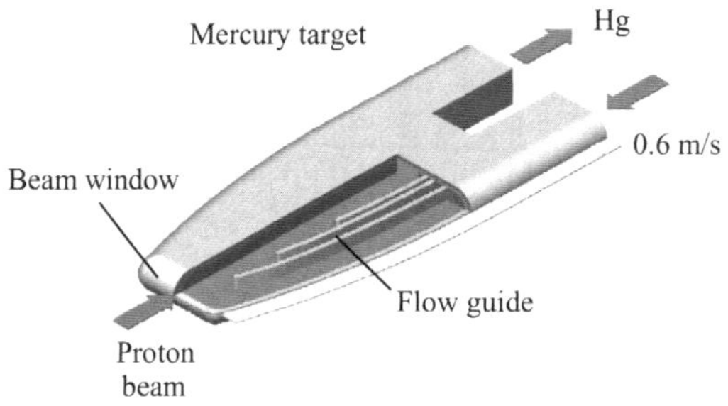

Figure 1. Mercury target vessel at the MLF/J-PARC.

Through theoretical and experimental investigations, the relationship between pressure wave conditions, aggressiveness of mercury cavitation, and damage growth behaviours in some solid metals can be understood. In order to mitigate the cavitation damage, we have applied two kinds of approaches: hardening of the vessel-wall and softening of the mercury. The former is achieved by coatings and/or surface improvements: Kolsterising®, nitriding, etc. [12]. The latter is enabled by the injection of micro-bubbles into the mercury, anticipating that the thermal energy caused by proton bombardment would be transformed to kinetic energy in the form of bubble oscillation [13, 14]. As for the structural materials investigation, the work carried out so far includes the very high cycle fatigue tests, the LME tests under mercury, irradiation effect evaluation tests using triple ions and the nano-indentation technique.

In this paper, research and development into techniques for the damage protection of the structure of the mercury target vessel is summarized.

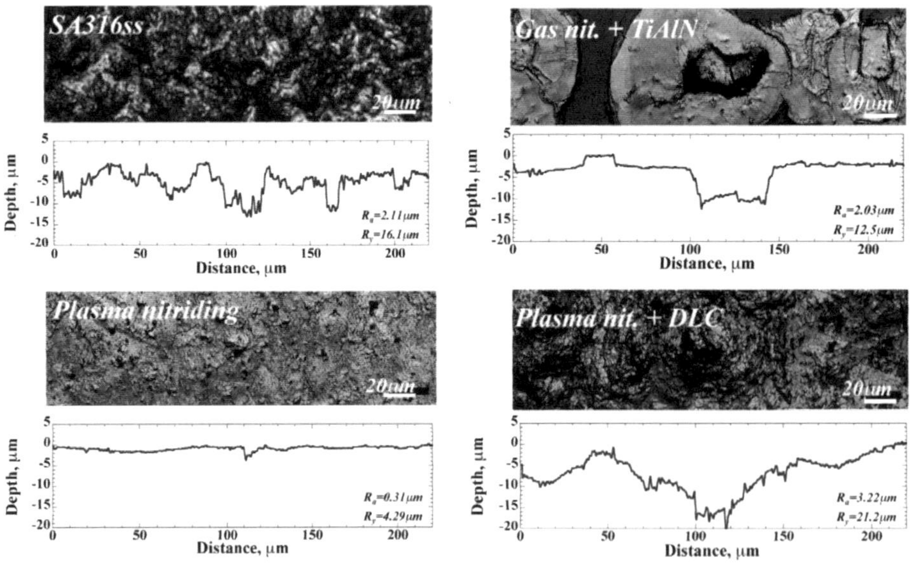

Figure 2. Typical examples of pitting damage surface protection treatments.

2. Mitigation of damage arising from cavitation

2.1. Hardening of vessel structural material

The high-cycle impact tests using an innovative electromagnetic impact machine, MIMTM, were carried out to understand the dynamic properties of mercury and the damage growth behaviour under localized impact due to cavitation in mercury [11].

More than 20 kinds of surface improvements: coating, implantation, etc. were examined experimentally to investigate suitable surface layer hardening protection against the cavitation erosion. The surface treatment was performed on solution-annealed austenitic stainless steel 316ss, which is the material used for the mercury target vessel. Figure 2 shows typical examples of pitting damage of samples with various surface treatments after imposing 10^6 impacts: these are optical images and the depth profiles are taken by a laser scanning microscope. The level of damage is highly dependent on the surface treatments, and the plasma nitriding treatment exhibits the most resistant surface against the pitting damage, nevertheless, the fatigue strength is degraded by the nitriding treatment, because of the critical property changes along the interface between improved surface and substrate. We developed, therefore, an innovative surface treatment consisting of plasma carburizing and nitriding, PCN, and the thickness of the improved surface was determined by the stress distribution along the interface between the substrate and the surface layer through impact analyses [15-17]. This treatment is expected to extend the duty period from 10^6 pulses to 10^7 pulses. This treatment was applied to the beam window of the first target vessel in the J-PARC.

2.2. Softening of target material; mercury

The injection of micro-bubbles into flowing mercury is expected to reduce the pressure waves and suppress the cavitation bubble growth [18-26]. The initial compressive pressure wave is reduced by the absorption of the thermal expansion of mercury due to the contraction of micro-bubbles, as shown in Figure 3 [18, 19]. The effect is expected at the region where the heat is deposited and at the onset of pressure wave propagation. During the wave propagation, the micro-bubbles reduce the amplitude of the compressive pressure waves through attenuation of the pressure waves, by thermal dissipation of kinetic energy and wave dispersion. These effects are highly dependent of the size and number of injected bubbles. In the mercury target, the anticipated bubble condition for mitigation is that the injected gas bubble size and void fraction are 50 μm and more than 10^{-4}, respectively [19].

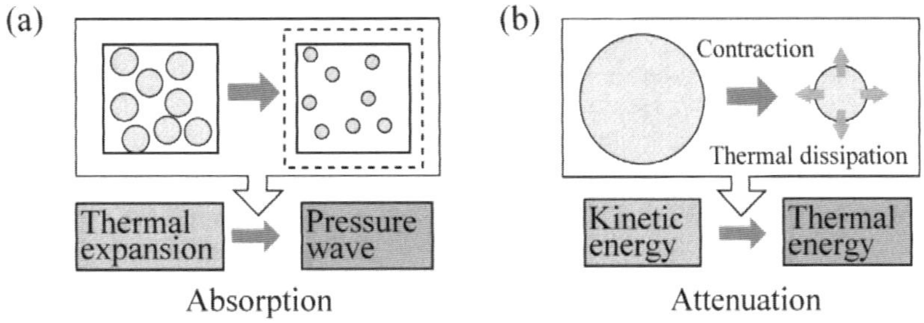

Figure 3. Softening mechanism of mercury against proton beam pressure waves; (a) absorption of the thermal expansion of mercury by the contraction of microbubbles, (b) attenuation of the pressure waves by the thermal dissipation of kinetic energy.

The two mechanisms provided a marked mitigation to the compressive pressure induced by the proton beam bombardment. In fact, the mitigation effect was recognized in the 2005 WNR on-beam test. The dynamic response of the mercury pipe wall was measured by using a laser Doppler vibrometer. The amplitude of velocity, which corresponds to the macroscopic pressure response caused by the proton beam bombardment, was clearly reduced by bubbling as shown in Figure 4 [25].

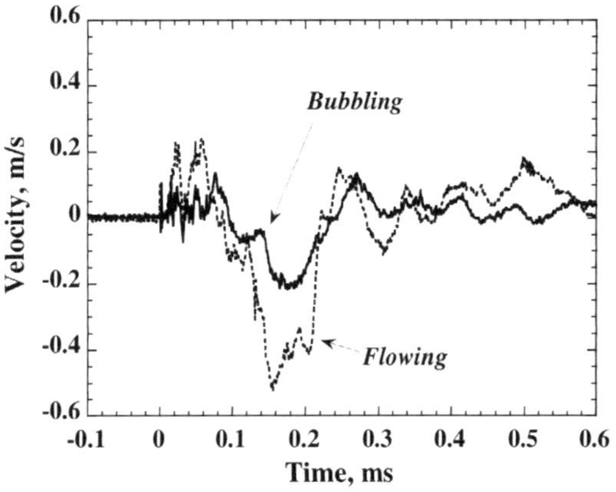

Figure 4. Displacement velocities measured at WNR on-beam tests.

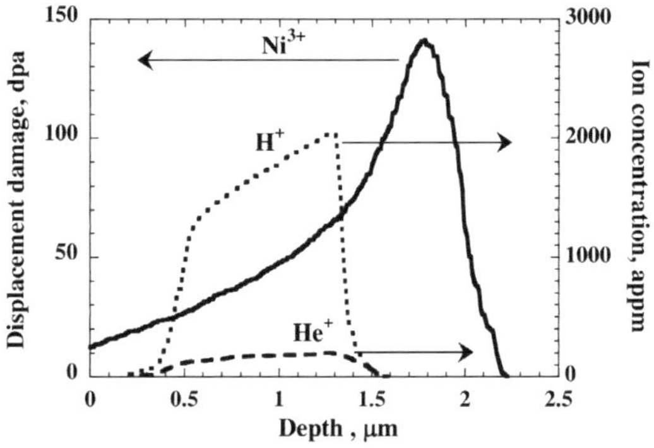

Figure 5. Depth profiles of displacement damage and implanted ion concentrations computed by SRIM code.

3. Irradiation damage and evaluation method for nano-scale zone

The indentation technique, for measuring the load-depth curve, is a very promising technology for the evaluation of mechanical properties, in particular for micro- and nano-meter scale zone of materials, including very thin layers. The novel technology uses the indentation technique combined with numerical calculation, and is developed in order to evaluate the mechanical properties quantitatively. The inverse analyses with the Kalman filter were carried out on the loading and unloading curves measured by the instrumented indentation machine with the hemispherical indenter [27, 28]. The technique is applicable not only to the quantitative evaluation of material constants in the micro- and nano-meter scale zones, including very thin surface layers, *i.e.* corroded layers, ion implanted layers, but also to the small specimen testing technique for irradiated materials.

The properties of the interface materials used as boundary against severe environments seem vary gradually with the distance from the surface. Material integrity might be influenced by the surface property degradation. Target vessel materials used in spallation neutron sources are exposed to proton and neutron irradiation and mercury immersion environments. In order to evaluate the surface degradation due to such environments, triple ion beam irradiation was carried out to the materials treated with surface hardening that are the candidate materials for a beam window in the target vessel. The mechanical properties of the irradiated gradient surface layer were evaluated by the indentation technique with inverse analysis, using multi-layer model [23].

In order to investigate the irradiation effect on the surface hardening treated 316ss, the candidate window material for the mercury target vessel, triple ion beam irradiation testing using H^+, He^+, Ni^{3+} was carried out to take account of the target spallation condition. Figure 5 shows the depth profiles of displacement damage and implanted ion concentrations, computed with SRIM code [29]. The changes of mechanical properties and microstructure due to irradiation were evaluated by the nano-indentation technique and TEM observation of the micro-specimen cut by FIB (Focused Ion Beam), as shown Figure 6 [23]. As a result, no significant change was observed on the surface hardening treated layers of 316ss. The surface hardening treatment was applicable to the target window materials to reduce the erosion damage.

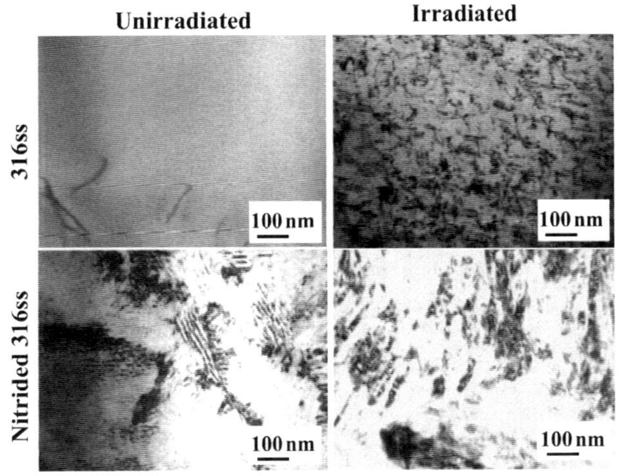

Figure 6. Cross-sectional TEM bright field images of un-irradiated and irradiated regions of 316ss and nitrided specimens. A few dislocation lines were observed in the un-irradiated 316ss, and larger dislocation loops with higher number density were formed in the irradiated 316ss. The martensitic structures with high dislocation density were observed in the un-irradiated nitriding, and there was hardly any change for the irradiated nitriding. It is considered that the dislocations of extremely high density in the martensitic structures are due to the nitriding suppressing the nucleation of radiation induced defect clusters by acting as a strong sink for point defects.

4. Liquid metal embrittlement

The effects of mercury immersion on fatigue crack propagation were investigated through three-point bending fatigue testing of notched specimens [30]. From the fatigue tests, in the low cycle fatigue region, the fatigue strength was seen to be slightly degraded by mercury immersion. In addition, the rate of increase in the measured notch opening distances of failed specimens with number of cycles, in the mercury case, was slightly larger than that for testing in air, regardless of the imposed stress, as shown Figure 7.

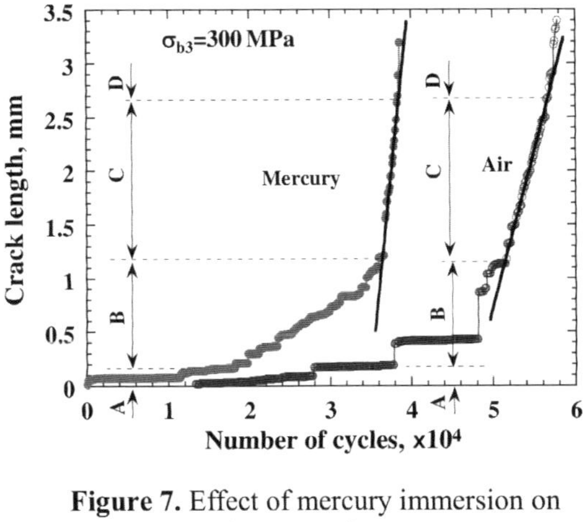

Figure 7. Effect of mercury immersion on crack propagation rates.

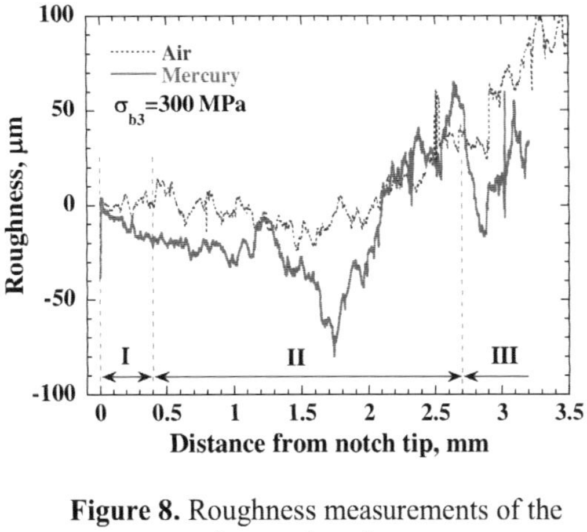

Figure 8. Roughness measurements of the fractured surfaces.

Figure 7 shows crack length as a function of the number of loading cycles. The rate of increase in the crack length as a function of the number of loading cycles can be classified as follows: incubation period A; a period, B, where the crack propagates gently according to the loading cycles; a period, C, where the crack propagates steadily and at a greater rate than during B; and finally, the fracture, at end point D. Region A, which corresponds to the early stage of Region I as shown in Figure 8, showed no differences between the mercury and air cases. In Region B, which the transitional region from Region I to II in Figure 8, a stepped crack length change was observed in the air case, which may be caused by unstable crack propagation. In Region C, a brittle fracture surface, *i.e.* an inter-granular fracture surface, was observed in the mercury case, as shown in Figure 9.

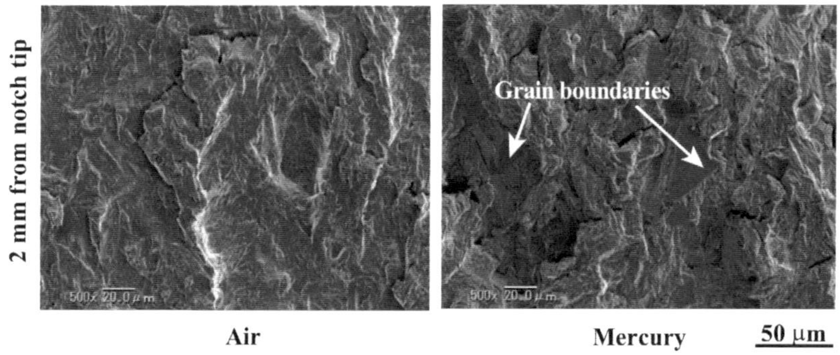

Figure 9. SEM images of the fracture surfaces at around 2 mm from the notch tip.

5. Summary

Impulsive pressure waves are a crucial obstacle to realizing the adoption of the mercury target for high power spallation neutron sources. High-cycle impact tests using an innovative electromagnetic impact machine, MIMTM, were performed to understand the dynamic properties of mercury, and the damage growth behaviour under localized impact due to cavitation in mercury. The hardening of the vessel surface and the injection of micro-bubbles, to soften the mercury and reduce the pressure wave amplitude, are expected to be useful for mitigating the localized impact damage resulting from cavitation. Additionally, the structural materials damage caused by proton and neutron irradiation, very high cycle fatigue and the so-called "liquid metal embrittlement" have been investigated through nano-indentation tests on triple-ion-implanted surface and crack propagation tests under mercury immersion. More detailed information on the research and development will be acquired from the references. As for very high cycle fatigue not described here because of space limitation, the recent progress is seen in ref. [31].

Acknowledgements

The author would like to thank all staff at the Neutron Source Section at MLF/J-PARC. This work was partially supported by the Japan Society for the Promotion of Science through a Grant-in-Aid for Scientific Research (Nos. 17360085, 17560740, 20360090 and 23360088).

References

[1] http://j-parc.jp/MatLife/en/ [last accessed October 2014].
[2] http://sns.gov/facilities/SNS/ [last accessed October 2014].
[3] Skala K, Bauer GS, 1995, *Proc. 13th Int. Advanced Neutron Sources (ICANS-XIII),* 559.
[4] Ni L, Bauer GS, Spitzer H, 1999, *Nucl. Instr. Meth. Phys. Res. A*, **425** 57.
[5] Futakawa M, Kikuchi K, Conrad H, Stechemesser H, 2000, *Nucl. Instr. Meth. Phys. Res. A*, **439** 1.

[6] Futakawa M, Kogawa H, Hino R, 2000, *J. Phys. IV France*, **10** 9-237.
[7] Futakawa M, Kogawa H, Hino R, Takeishi H, 2003, *Int. J. Impact Eng.*, **28** 123.
[8] Haines JR, Riemer BW, Felde DK, Hunn JD, Pawel SJ, Tsai CC, 2005, *J. Nucl. Mater.*, **343** 58.
[9] Futakawa M, Naoe T, Tsai C, Kogawa H, Ishikura S, Ikeda Y, Soyama H, Date H, 2005, *J. Nucl. Mater.*, **343** 70.
[10] McClintock D, Riemer B, Ferguson P, Carroll A, Dayton M, 2012, *J. Nucl. Mater.* **431** 147.
[11] Futakawa M, Naoe T, Kogawa H, Tsai C, Ikeda Y, 2003, *J. Nucl. Sci. Technol.,* **40** 895.
[12] Mansur , 2003, *J. Nucl. Mater.,* **318** 14.
[13] Farrell K, Specht ED, Pang J, Walker LR, Rar A, Mayotte JR, 2005, *J. Nucle. Mater.* **343** 123.
[14] Naoe T, Futakawa M, Shobu T, Wakui T, Kogawa H, Takeuchi H, Kawai M, 2008, *J. Nucl. Sci. Technol.* **45** 698.
[15] Futakawa M, Kogawa H, Hasegawa S, Ikeda Y, Riemer B, Wendel M, Haines J, Bauer G, Naoe T, Okita K, Fujiwara A, Matsumoto Y, Tanaka N, 2008, *J. Nucl. Mater.* **377** 1828.
[16] Futakawa M, Naoe T, Kogawa H, Wakui T, Haga K, 2011, *Proc. WIMRC 3rd Int. Cavitation Forum*, July, University of Warwick, Coventry, UK.
[17] Naoe T, Futakawa M, Shobu T, Wakui T, Kogawa H, Takeuchi H, Kawai M, 2008, *J. Nucl. Sci. Technol.,* **45** 698.
[18] Soltner H, 2003, *Gas bubble admixture for pressure pulse mitigation in high-power liquid-mercury spallation targets*, ISSN 1433-559X, ESS03-152-T.
[19] Okita K, Takagi S, Matsumoto Y, 2008, *J. Fluid Sci. Technol.,* **3** 116,
[20] Futakawa M, Kogawa H, Hasegawa S, Naoe T, Ida M, Haga K, Wakui T, Tanaka N, Matsumoto Y, Ikeda Y, 2008, *J. Nucl. Sci. Technol.*, **45** 1041.
[21] Naoe T, Ida M, Futakwa M, 2008, *Nucl. Instr. Meth. Phys. Res. A,* **586** 865.
[22] Futakawa M, Naoe T, Kogawa H, Date H, Ikeda Y, 2005, *JSME Int. J.,* **A48** 234.
[23] Naoe T, Futakawa M, Naito A, Kogawa H, Ikeda Y, Motohashi Y, 2005, *JSME Int. J.,* **A48** 280.
[24] Naoe T, Kogawa H, Futakawa M, Ida M, 2011, *J. Nucl. Sci. Technol.,* **48** 865.
[25] Futakawa M, Naoe T, 2011, *Exp. Therm. Fluid Sci.*, **35** 1177.
[26] Kogawa H, Haga K, Naoe T, Kinoshita H, Ida M, Futakawa M, 2010, *Proc. 19th Int Advanced Neutron Sources (ICANS-XIX),* March, Grindelwald, Switzerland.
[27] Futakawa M, Ioka I, Wakui T, Tanabe Y, 2001, *Proc. 10th APCNDT,* Brisbane.
[28] Futakawa M, Wakui T, Kogawa H, Date H, 2002, *Proc. JSME/ASME Int. Conf. Mat. Proc.,* Honolulu, 558.
[29] Ziegler JF, Biersack JP, Littmark U, 1985, *The stopping and tange of ions in solid,* Pergamon Press, New York.
[30] Naoe T, Yamaguchi Y, Futakawa M, 2012, *J. Nucle. Mater.,* **431** 133.
[31] Zhihong X, Futakawa M, Naoe T, Maekawa K, 2014, *Adv. Mater. Res.,* **891-892** 536.

IOP Conf. Series: Materials Science and Engineering **74** (2015) 012002 doi:10.1088/1757-899X/74/1/012002

Linear and non-linear electrical dependency of carbon nanotube reinforced composites to internal damage

D Bekas[1], **SA Grammatikos**[2], **C Kouimtzi**[1], **AS Paipetis**[1]

[1] Dept. of Materials Science & Engineering, University of Ioannina, 45110 Ioannina, Greece.
[2] Dept. of Architecture and Civil Engineering, University of Bath, Bath BA27AY, UK.

paipetis@cc.uoi.gr

Abstract. Carbon nanotube (CNT) enhanced composite materials have attracted the interest of many scientists worldwide, especially in the aerospace industry. Fundamental to their qualification as materials in primary aircraft structures is the investigation of the relationship between their functional characteristics and their long-term behaviour under external combined loads. Conductive reinforcement at the nanoscale is by definition multifunctional as it may *(i)* enhance structural performance and *(ii)* provide structural health monitoring functionalities. It is now well established that reversible changes in the electrical resistance in nano composites are related to strain and irreversible monotonic changes are related to cumulative damage in the nano composite. In this study, the effect of damage in the hysteretic electrical behaviour of nano-enhanced reinforced composites was investigated. The nanocomposites were subjected to different levels of damage and their response to a cyclic electrical potential excitation was monitored as a function of frequency. Along with the dynamic electrical investigation, an Electrical Potential Mapping (EPM) technique was developed to pin-point artificial damage in CNT-enhanced matrix composite materials. The electrical potential field of the bulk material has shown to be characteristic of its internal structural state. The results of EPM technique were contradicted and validated with conventional C-scans.

1. Introduction

Carbon nanotubes (CNTs) possess a unique combination of mechanical, electrical, and thermal properties that makes them excellent candidates to substitute or complement the conventional nano-fillers in the fabrication of multifunctional polymer nano-composites [1-6], mainly in the field of aerospace industry. During the past decade, epoxy resin-based CNT composites were extensively studied by many researchers for their electrical properties [7-11]. It was well established that, these composites could potentially be used for transparent conductive coatings, electrostatic dissipation, electrostatic painting and electromagnetic interference shielding applications..

When employed as an indicator for structural integrity of a component; the electrical properties of composite materials are capable of achieving a direct correlation between the internal damage and the inherent properties. Many researchers managed to relate irreversible electrical resistance change to internal damage of the composite material [12-17]. The structural integrity of Carbon Fiber Reinforced Polymers (CFRPs) with Multi Wall Carbon NanoTubes (MWCNTs) in the matrix material was evaluated concurrently with the self-sensing abilities of the hybrid composite [18].

Self-sensing is achieved via the employment of the percolated electrical network within the matrix which performs as a sensor in the structure. The changes in the electrical properties of nano-composites and CFRPs, when subjected to load provide a direct measure for both strain and damage accumulation.

On the other hand, dielectric based methodologies have become one of the most interesting techniques for the on-line monitoring of changes that occur to physical and chemical properties of polymeric resins and their composites during the curing process. Through dielectric analysis the instantaneous degree of cure, cure rate, and viscosity evolution have been predicted [19-21]. Recently, the CNT dispersion process in epoxies has been related to the impedance of the system and has been successfully modeled as an equivalent RLC circuit [22, 23], however, little research effort has been devoted to the study of the impedance properties of nano-reinforcement composites and its relation to loading history and/or internal damage. Within the scope of this work is to enhance the resolving ability of the electrical based methodologies via the employment of AC configurations. These involve the study of the effect of damage in the hysteretic electrical behavior of carbon nanotube reinforced epoxy nano composites. In a parallel study, an Electrical Potential Mapping (EPM) technique was developed to pin-point artificial damage in CNT-enhanced matrix composite materials using DC resistance measurements, with a view to the development of an automated impedance mapping facility and its correlation to damage evolution in the structure.

2. Experimental

2.1. Materials

For the EPM (DC measurements), Carbon nanotube modified epoxy resin CFRPs were manufactured. 16-layer quasi-isotropic CFRPs with 0.5% w/w CNT modified epoxy matrix were manufactured. The first epoxy matrix used was the Araldite LY 5052/Aradur 5052 by Huntsman International LLC (Switzerland). Multi Wall Carbon Nanotubes (MWCNTs) by Arkema, France, were employed as additives in the epoxy matrix of the hybrid composites. MWCNTs were synthesized by Catalyzed Chemical Vapor Deposition (CCVD) and the tube diameter ranged from 10 to 15 nm, the tube length was more than 500 nm. CFRP laminated were produced using the infusion method. After curing, 60mmx60mm square specimens were cut from the laminates.

Figure 1. Sonication process.

For the hysteretic investigation (AC measurements), the Araldite LY 564 and Aradur 2954 by Huntsman Advanced Materials, Switzerland was employed for matrix. As for the reinforcement, MWCNTs were used. CNTs were identical to the EPM concept. In all cases, good dispersion quality of the CNTs in the epoxy matrix was achieved via sonication.

2.2. Processing of the nanocomposites: Sonication

In order to disperse the CNTs into the epoxy resin an ultrasonic mixer (UP400S, Hielscher) was used. CNTs and resin were carefully weighed and mixed together in a beaker. An ice bath was also used so as to avoid overheating of the polymer resin and introduction of defects on the CNTs surface,

Figure 1. CNTs were dispersed in the matrix using sonication for 4 hrs at 100% amplitude. At the end of this process the hardener was added to the modified resin and mixed using a mechanical agitator for about 10 minutes. At the end, the first mixture (Araldite LY 5052/Aradur 5052) transferred into silicon rubber molds and cured at 23 °C for one day and the second mixture (Araldite LY 564/ Aradur 2954) cured at 60 °C for 2 hours. Post curing of first and second systems was carried out at 100 °C for 4 hours and 120 °C for 4 hours respectively. It must be pointed out that the addition and the dispersion of the CNTs into the polymeric matrix resulted to an increase of the glass transition temperature (Tg) of the polymer. This phenomenon can be attributed to the interaction between the CNTs and the epoxy matrix [24].

2.3. Testing procedure

2.3.1. Impact damage/drilled open hole. For the EPM study, two damage concepts were investigated; impact damage and drilled hole. A 5 mm drilled hole, a 3 J and a 5 J blind impact damage were inspected. The circular notch was drilled using a diamond drill in the centre of the specimen. Dynamic impact tests were carried out using an Instron (CEAST 9340) drop-weight round impactor. Before the impact, the specimens were clamped on the testing machine. The performed impact energy levels did not lead to penetration of the employed composite laminates. In all cases, an Agilent multimeter was employed for the electrical resistance and electrical potential measurements. Regarding the resistance technique, the two-wire approach was adopted. For the electrical potential mapping, 100mA electrical current was injected by an external DC power supply.

2.3.2. Cycling Thermal Shock. Cyclic thermal shock was performed in order to simulate temperature changes which occur to aircraft structures. The specimens divided into four groups. The first group was undamaged and subsequently used as reference. Second, third and fourth group of samples subjected to a range of 30, 50 and 70 cycles, respectively. The temperature changes ranged from -30 °C to +30 °C. Samples remained at these temperatures for one hour.

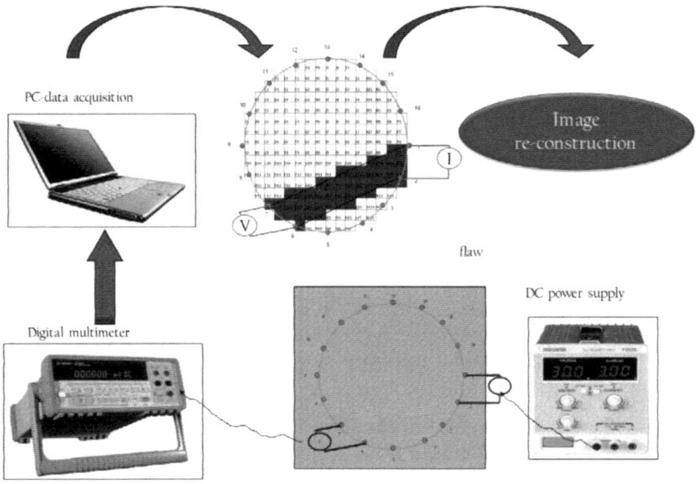

Figure 2. Experimental setup for EPM surface measurement.

2.3.3. Electrical Potential Mapping (EPM) system implementation. In Figure 2 the EPM experimental setup and the measurement protocol are depicted. 150 mA DC current was injected through an electrical contact pair. The voltage between all the remaining electrical adjacent contact pairs was subsequently measured using the digital multimeter. The same measurement sequence was followed by moving the pair of the injecting electrodes and measuring the voltage at all the respective remaining electrical contacts.

As is shown in Figure 2, the electrical contacts on the surface of the investigated coupons draw a circle which was segmented in separated cells. In specific, a square of 16 × 16 cells was formed in excel. Every cell represents an average of electrical potential measurements. Further on, the two lines which connect the input and output pair of electrodes in the grid representation define the cells that will be include in the average of each cell. According to the employed protocol, when more than one third of the cell is included in between the lines, then this cell is included in the total average measurements.

2.3.4. Impedance (AC) measurements. An instrumentation of an AC power supply source (dual channel arbitrary/function generator AFG 3052C) and a two channel digital storage oscilloscope (TDS 2002C) supplied by Tektronix was developed, Figure 3. Silver conductive paint and silver conductive tape were used to achieve good electrical coupling between the sample surface and the electrodes.

In order to obtain the hysteresis curves, the power supply created a voltage waveform with constant amplitude of 21 Volts and a frequency that varied from 80 Hz to 2 kHz through the examined material. On the other hand oscilloscope's channel 1 and channel 2 recorded the inbound and outbound voltage waveform respectively. Then the two waveforms are plotted in the same graph in oscilloscope's screen. Finally phase-φ and amplitude from peak to peak were also displayed.

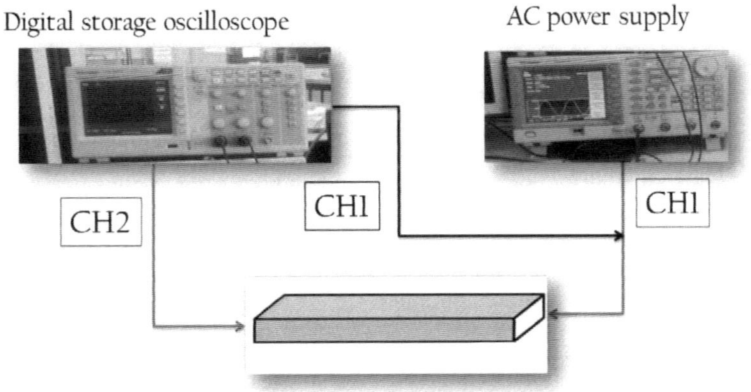

Figure 3. Experimental setup for the electrical (AC) measurements.

3. Results and Discussion

3.1. Electrical Potential Mapping (EPM)-DC measurements

Figure 4 depicts the damaged material with the 5 mm drilled hole. The differentiation in colour in the centre of the image represents the presence of the artificially induced drilled hole in the centre of the specimen.

Figure 5 displays the impacted plate against the employed NDE methods for the 3 J impact, which indicates that the damage is visible in the form of a semi-spherical dent at the impact site. As can be seen in Figure 5b, the induced damage by the round impactor is within the resolving power of the C-scan image. The exhibited color differentiation on the image indicates that measurements identify damage in the center of the specimen.

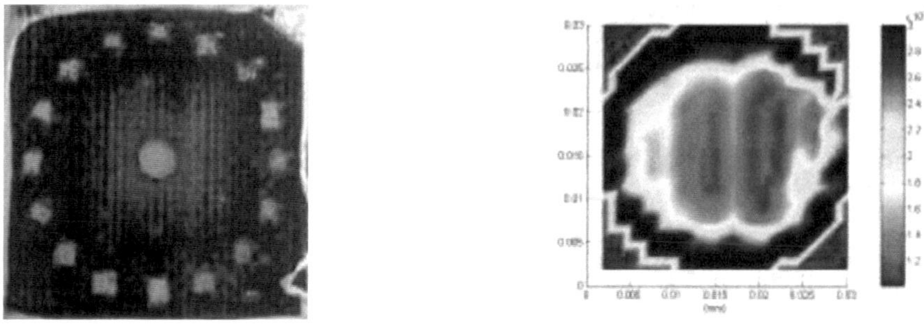

(a) (b)

Figure 4. (a) Cross-ply CFRP-5mm drilled hole (b) EPM image.

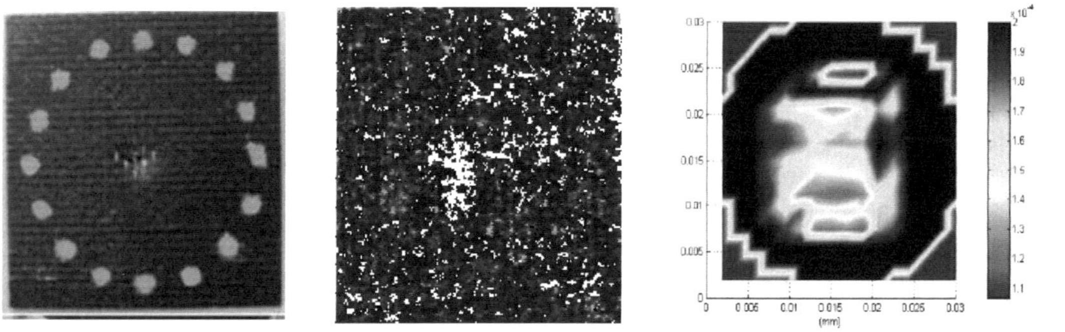

(a) (b) (c)

Figure 5. (a) Cross-ply a 3J impact damage (b) C-scan (c) EPM image.

In Figure 6a the 5 J LVI damaged specimen is presented. In this case, damage is barely or invisible to the naked eye. Figure 6b shows the caused impact damage through an amplitude C-scan image. Figure 6c shows the EPM image. Concerning the EPM image, the deterioration of the material is expected to be identified in the centre based on the ultrasonic testing.

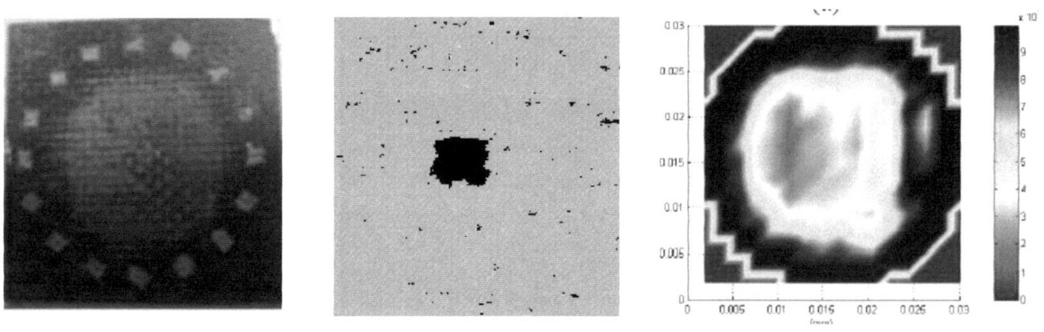

(a) (b) (c)

Figure 6. (a) Cross-ply a 5 J impact damage, (b) C-scan (c) EPM image.

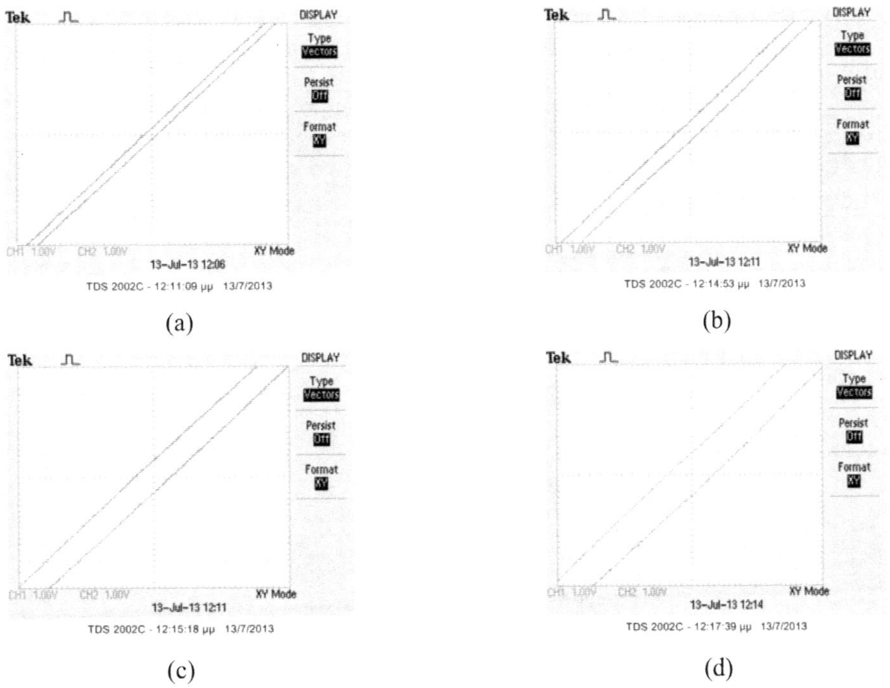

Figure 7. V-V curves for the specimen subjected to thermal shock (30 cycles) at: (a) 250Hz of AC voltage; (b) 450 Hz of AC voltage; (c) 750 Hz of AC voltage; (d) 1 kHz of AC voltage.

3.2. AC measurements

Four V-V curves were generated at 250, 450, 750 and 1000 Hz for every sample. Figure 7 presents the hysteretic electrical behavior of carbon nanotube reinforced epoxy nanocomposite that subjected to 30 cycles of thermal shock. As long as the material is under an electric field, CNTs act like dipoles. These dipoles are trying to orient to the direction of the electric field. As frequency of the AC increases dipoles cannot follow these changes properly, thus, wider hysteresis cycles can be observed. Moreover, changes in hysteresis cycles observed for the same frequency of AC voltage in different levels of damage. This becomes more obvious in the plots, Figure 8, where the V-V curves with 1 kHz for the four specimens are presented.

The measured phase delay (-) is increasing both with frequency and the level of damage for all observed systems. Phase delays are recorded for frequencies above approximately 200 Hz. This can be clearly distinguished in Figure 9 where phase-φ – frequency graphs are depicted for every sample with different levels of damage. Specimens subjected to 50 and 70 cycles of thermal shock show the largest difference in both in phase and drop rate of the phase delay when compared to the reference one.

4. Conclusions

The scope of the work described in this section was to examine the linear and non-linear electrical dependency of nano-composites to internal damage. Initially, an electrical potential mapping (EPM) system was developed with a view to identifying induced damage in composite laminates.

The implementation of the EPM technique proved its reliability and effectiveness in identifying both the notch and the impact induced damages in the laminates. The EPM algorithm identified potential differences corresponding to the location of the drilled-hole and blind impact damage, moreover, the enhancement of the resolving ability of the electrical based methodologies via the employment of AC configurations was studied.

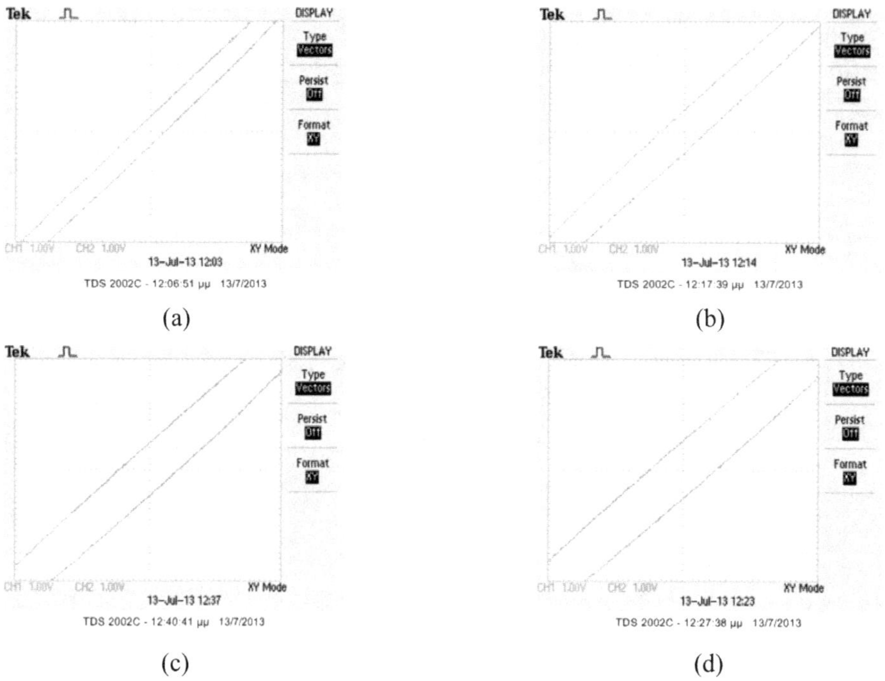

Figure 8. V-V curves generated using AC voltage with 1 kHz frequency for specimens subjected to thermal shock: (a) 0 cycles; (b) 30 cycles; (c) 50 cycles; (d) 70 cycles.

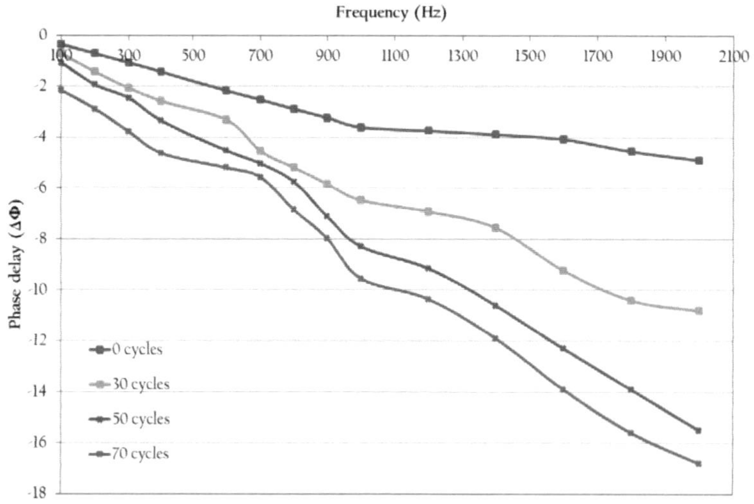

Figure 9. Phase-φ – frequency curves for four samples with different levels of damage.

The effect of damage in the hysteretic electrical behaviour of nano-enhanced reinforced composites was investigated. The results obtained from the AC voltage measurements indicate that the hysteresis curves become wider as the level of damage increases and a correlation between phase-φ and frequency of the AC voltage was achieved. Currently effort is being devoted in combining the aforementioned methodologies (AC and DC) for the accomplishment of a fully automated impedance mapping in order to increase the accuracy of the solution of the inverse problem by introducing a higher dimension to the problem (both measure and phase).

The implementation of the EPM technique proved its reliability and effectiveness in identifying both the notch and the impact induced damages in the laminates. The EPM algorithm identified potential differences corresponding to the location of the drilled-hole and blind impact damage, moreover, the enhancement of the resolving ability of the electrical based methodologies via the employment of AC configurations was studied.

The effect of damage in the hysteretic electrical behaviour of nano-enhanced reinforced composites was investigated. The results obtained from the AC voltage measurements indicate that the hysteresis curves become wider as the level of damage increases and a correlation between phase-φ and frequency of the AC voltage was achieved. Currently effort is being devoted in combining the aforementioned methodologies (AC and DC) for the accomplishment of a fully automated impedance mapping in order to increase the accuracy of the solution of the inverse problem by introducing a higher dimension to the problem (both measure and phase).

References

[1] Bauhofer W, Kovacs JZ, 2009, *Composites Science and Technology* **69** 1486–1498.
[2] Qian H, Greenhalgh ES, Shaffer MSP, Bismarck A, 2010, J. Mater. Chem 20 4751–4762.
[3] Rahmat M, Hubert P, 2011, *Composites Science and Technology* **72** 72–84.
[4] Ma PC, Siddiqui NA, Marom G, Kim JK 2010, *Composites: Part A* **41** 1345–1367.
[5] Spitalsky Z, Tasis D, Papagelis K, Galiotis C, 2010, *Progress in Polymer Science* **35** 357–401.
[6] Allaoui A, Bai S, Cheng HM, Bai JB, 2002, *Composites Science and Technology* **62** 1993–1998.
[7] Gojny FH, Wichmann MHG, Fiedler B, Kinloch IA, Bauhofer W, Windle AH, Schulte K, 2006, *Polymer* **47** 2036–2045.
[8] JSandler J, Shaffer MSP, Prasse T, Bauhofer W, Schulte K, Windle AH, 1999, *Polymer* **40** 5967–5971.
[9] Huang Y, Li N, Ma Y, Du F, Li F, He X, Lin X, Gao H, Chen Y, 2007, *Carbon* **45** 1614–1621.
[10] Zheng W, Wong SC, 2003, *Composites Science and Technology* **63** 225–235.
[11] Sandler JKW, Kirk JE, Kinloch IA, Shaffer MSP, Windle AH, 2003, *Polymer* **44** 5893–5899.
[12] Shindo Y, Kuronuma Y, Takeda T, Narita F, Fu SY, 2012, *Composites Part B* **43** 39-43.
[13] Fiedler B, Gojny FH, Wichmann MHG, Bauhofer W, Schulte K, 2004, *Sci Mat* **29** 81–94.
[14] Nofar M, Hoa SV, Pugh MD, 2009, *Compos Sci Technol* **69** 1599–1606.
[15] Thostenson ET, Chou TW, 2006, *Adv Mater* **18** 2837–2841.
[16] Park M, Kim H, Youngblood JP, 2008, *Nanotechnology* **19** 055705 (7pp).
[17] Kang I, Schulz MJ, Kim JH, Shanov V, Shi D, 2006, *Smart Mater. Struct.* **15** 737–748.
[18] Grammatikos SA, Paipetis AS, 2012, *Composites: Part B* **43** 2687–2696.
[19] Kim D, Centea T, Nutt SR, 2014, *Composites Science and Technology* **102** 132–138.
[20] Maistros GM, Partridge IK, 1998, *Composites: Part B* **29B** 245–250.
[21] McIlhagger A, Brown D, Hill B, 2000, *Composites: Part A* **31** 1373–1381.
[22] Gkikas G, Saganas C, Grammatikos SA, Maistros GM, Barkoula NM, Paipetis AS, 2012, *Sensors and Smart Structures Technologies for Civil, Mechanical, and Aerospace Systems.*
[23] Gkikas G, Saganas C, Grammatikos SA, Aggelis DG, Paipetis AS, 2012, *Smart Sensor Phenomena, Technology, Networks, and Systems Integration* (7pp).
[24] Gkikas G., Barkoula N.-M., Paipetis A. S. 2012 *Composites: Part B* **43** 2697–2705.

Spring back of infinite honeycomb sheets beyond plastic deformation

A Bonfanti and A Bhaskar

Computational Engineering and Design Group, Faculty of Engineering and the Environment, University of Southampton, UK

E-mail: ab38g12@soton.ac.uk

Abstract. Cellular structures are promising for applications where high stiffness and strength are required with the minimal use of material. They are often used in applications where the plastic deformation plays an important role, such as those involving crashworthiness, energy absorption, and stents. The elastic analysis of a honeycomb sheet has been carried out in the past [1]. The present analysis extends this classical work in the elasto-plastic regime. Recoil analysis due to elastic recovery is absent from the published literature. This work aims to develop an analytical model to calculate the spring back for a simplified case, that of an infinite honeycomb sheet. An elastic-perfectly plastic material model is assumed. The recoil for a clamped beam with a load and moment applied at the free edge is analytically calculated first. This is carried out by relating the stress distribution of the cross section to the final deformed shape. The part corresponding to the elastic contribution is subsequently subtracted in order to obtain the final configuration after the external load is removed. This simple elasto-plastic analysis is then incorporated into the analysis of an infinite sheet made of uniform hexagonal cells. The translational symmetry of the lattice is exploited along with the analysis of a beam under tip loading through to plastic stage and recoil. The final shape of the struts upon the removal of the remote stress is completely determined by the plastic deformation which cannot be recovered. The expression for the beam thus obtained is then used to build an analytical model for an infinite honeycomb sheet loaded in both directions.

1. Introduction

Periodic cellular architectures are frequently found in several natural material and engineered structures. Synthetic structured materials can be found in sandwich panels, heat exchangers, cushioning foams and biological implants. In the last twenty years, they have become an important class of material since they afford light weight structures with high stiffness and strength. By minimizing the weight of an engineering system, its performance can be greatly improved such as in aircraft construction where the core of sandwich panels is made of honeycomb sheets [2]. On the other hand, by minimizing the material employed in biological implants, side effects can be drastically decreased, such as the restenosis in stenting procedure used to treat cardiovascular arterial obstructions [3].

The mechanical properties of cellular solids depend on the constituent material and the cellular architecture (relative density and cell shape). Based on the cell shape, several classifications have been made. The first is the distinction between two-dimensional and three-dimensional cells (often called "foam"). The analysis of two-dimensional cellular solids is simpler, therefore the mechanical properties can be analytically related to the parameters that describe the unit cell structure by using Euler-Bernoulli beam analysis [2]. Further, the cellular geometry can be classified on the basis of the presence of translational symmetry in case of lattice material vs. disordered cellular solids that lack this attribute (see figure 1). The hexagonal lattice is perhaps the most frequently studied ordered structure – the regularity coupled with the apparent isotropy of this lattice being the important reasons for the great interest amongst physicists, engineers and mathematicians [1].

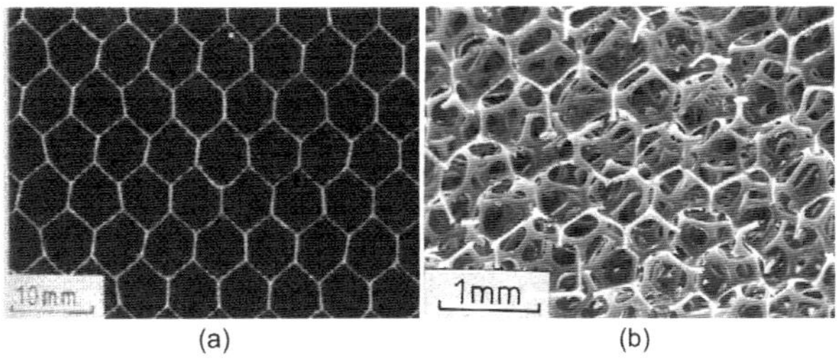

Figure 1. An example of (a) aluminium honeycomb sheet and (b) a disordered polyurethane cellular material [1].

Many engineered products and material such as cores of sandwich panels for doors or aerospace components, heat exchangers and crash absorbers possess hexagonal internal structure. Secondly, several generic properties that characterise two-dimensional cellular solids can be extended to three-dimensional geometries. Due to their geometrical regularity, the apparent mechanical properties can be easily obtained in a closed form [1]. Most work in the past has been limited to linear-elastic analysis of such structures with the exception of the collapse using a plastic hinge analysis [1]. On the other hand, many applications require detailed elasto-plastic analysis, including the calculation of elastic response followed by plastic deformation. When a plastically deformed material is unloaded, the elastic deformation is recovered leading to recoil or spring back. The work of Ashby and Gibson provided the apparent Young's modulus, Poisson's ratio and collapse pressure for the cellular material by using defined equations. However, plastic analysis and recoil calculations for such lattices are missing in the literature.

The purpose of this study is to extend the work by Gibson and Ashby [1] by including an analytical model that can capture the recoil for two dimensional infinite honeycomb elastic lattices. Given that the load is applied at infinity, the apparent constitutive model must provide the spring back of such structured sheets.

2. The elasto-plastic bending and spring back of an infinite honeycomb sheet subjected to uniaxial remote force

Consider a regular two-dimensional cellular hexagonal honeycomb as the one shown in figure 2(a). The honeycomb geometry is kinematically mobile, therefore the cell walls deform primarily in flexure. It is common to refer to this type of structure as bending-dominated [4]. The constitutive relationship used here is the elastic-perfectly plastic idealisation of the material behaviour.

By using thin beam theory, expressions for the deflection of the cell wall under elastic deformation have been developed in the past [1]. Even though flexure is the dominant mechanism, the stretching influences the structural response. Therefore, this model has been extended by introducing the effects of the axial force [1]. Since this model is valid only for the elastic deformation of the honeycomb cell, it is not valid when the stress exceeds the yield stress. The aim of this study is to provide an analysis valid also when the honeycomb sheet is subjected

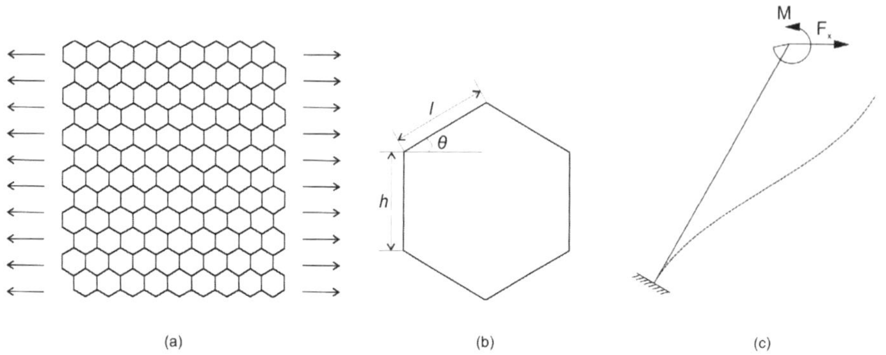

Figure 2. (a) honeycomb sheet, (b) undeformed honeycomb parameters and (c) cell wall deformation under horizontal remote stress.

to plastic deformation. The calculation of the beam deflection has been carried out by relating it to the stress distribution of the cross section by using the fundamental relationship between curvature and load of a cantilever beam provided by [5].

Consider the honeycomb sheet shown in figure 2(a). When a remote uniaxial stress is applied along one of the principal directions, the force experienced at the cell wall end is shown in figure 2 (c).

The resulting moment distribution of the cell wall is shown in figure 3. The mechanics of elasto-plastic deformation of beams has been studied by Yu and Johnson [5] who considered a cantilevered beam under tip load in the context of predicting spring back in the die-punch problem. We have used this analysis by exploiting the symmetry and using the outcome of [5]. As a result, the deformation of the cell wall is obtained. As expected, the deflected shape is cubic, which is obtained by integrating a fourth order equilibrium equation. By repeating this solution symmetrically for the four inclined cell walls, the final cell shape under remote stress can be obtained for the plastic phase first and then the recoil.

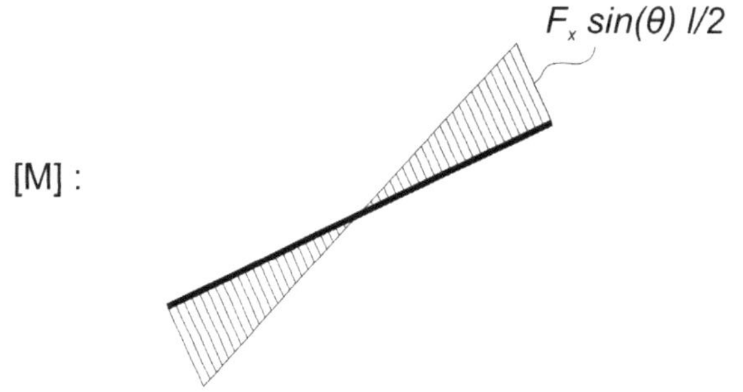

Figure 3. Moment distribution along the cell wall under horizontal remote stress.

Now the elastic spring back after removing the horizontal stress applied at infinity is considered. According to [5], the unloading is obtained by superimposing the elastic effects caused by the application of a reverse force at the free end of the beam. The resulting qualitative shape of the cell wall is shown in figure 4.

IOP Conf. Series: Materials Science and Engineering **74** (2015) 012003 doi:10.1088/1757-899X/74/1/012003

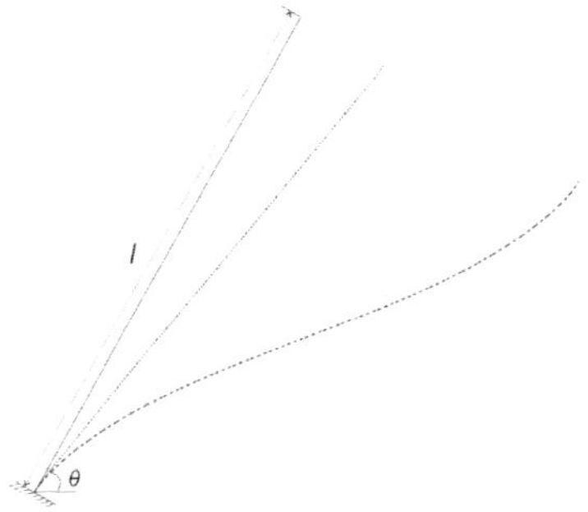

Figure 4. The deflection of the cell wall: — the elasto-plastic deflection, ... the final shape after unloading.

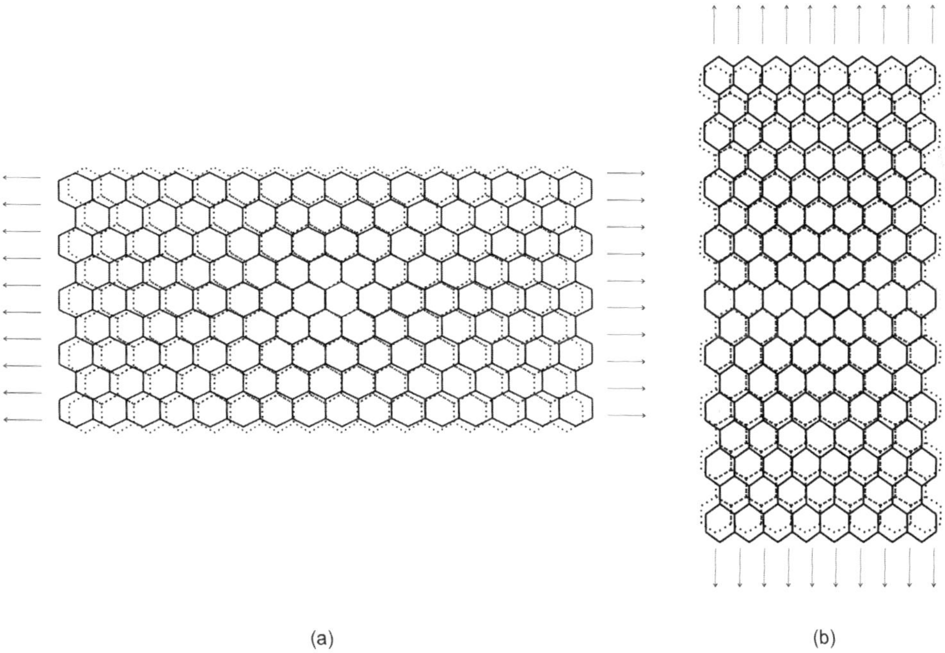

Figure 5. Elasto-plastic response of an infinite honeycomb sheet subjected to (a) horizontal remote stress and (b) perpendicular remote stress.

3. Results

By using the analytical model described in the previous section, an elasto-plastic analysis followed by the elastic spring back analysis of an infinite honeycomb sheet was carried out. Consider the infinite honeycomb sheet in figure 2 (a) subjected to a horizontal stress. Figure 5 (a) shows the final shape obtained after the application of the load. The same analysis has been performed by applying the remote stress along direction perpendicular to the first. These results are plotted in figure 5 (b). As can be noticed, the sheet shrinks in the perpendicular direction of the force application and elongates in the other due to the Poisson's ratio effect of the porous structure. The amount of this lateral response has been exactly calculated here using the analytical model described in the previous section.

Consider the same infinite honeycomb sheet subjected to remote stress. Assume that this sheet is now unloaded. When the elastic deformation is recovered, the only deformation left on the sample is the one related to the permanent plasticity. Such mechanical response is often known as spring back or recoil. Figure 6 shows the final deformed shape of the sheet. To the best of our knowledge, it is the first time that spring back of infinite lattice structures has been reported.

This model can analytically predict the recoil of an infinite honeycomb sheet. Even though hexagonal shape is the most common, several other shape geometries can be found in real life. Therefore, the next step will involve the extension of this approach to more general geometries.

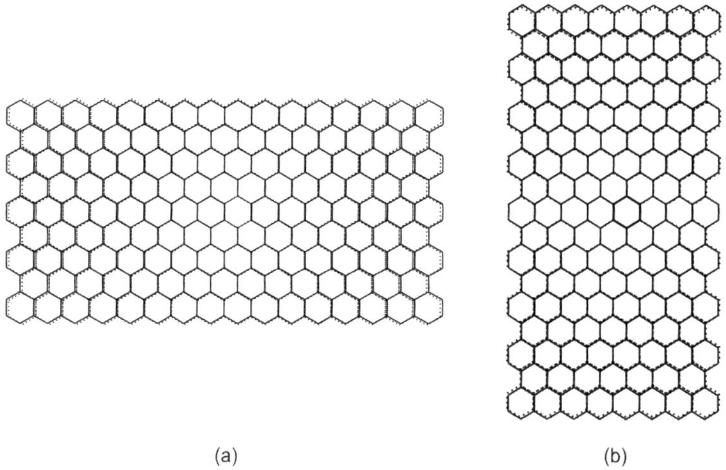

Figure 6. Springback of an infinite honycomb sheet subjected to (a) horizontal remote stress and (b) perpendicular remote stress after the unloading.

4. Conclusions

Cellular materials are nowadays applied in several applications which require plastic deformation. An inevitable effect is the spring back or recoil. This needs to be accounted for in the analysis so that the shape of the structure upon removing external loading is what intended by the designer. Using a simple mechanical approach which relates the curvature to the cross section stress distribution, we have carried out the final deformation under elasto-plastic conditions. By subtracting the elastic effects from the section stress distribution, the final curvature after the unload has been calculated. One of the advantages of the proposed approach is the analytical expressions for various response quantities as opposed to computational methods such as the Finite Element method which does not provide an explicit parametric dependence of the results. Further details will be published elsewhere [6].

References

[1] Gibson L J and Ashby M F 1999 *Cellular solids: structure and properties* (Cambridge university press)
[2] Abd El-Sayed F, Jones R and Burgess I 1979 *Composites* **10** 209–214
[3] Gunn J and Cumberland D 1999 *European heart journal* **20** 1009–1013
[4] Bhaskar A 2009 *EPL (Europhysics Letters)* **87** 18004
[5] Yu T and Johnson W 1982 *Journal of Mechanical Working Technology* **6** 5–21
[6] Bonfanti A and Bhaskar A 2014 *[Manuscript in preparation]*

Mechanical characterisation of additively manufactured material having lattice microstructure

E Cuan-Urquizo[1], **S Yang**[2] **and A Bhaskar**[1]

[1] Computational Engineering and Design Group, Faculty of Engineering and the Environment, University of Southampton, SO17 1BJ, UK

[2] Engineering Materials Research Group, Faculty of Engineering and the Environment, University of Southampton, SO17 1BJ, UK

E-mail: E.CuanUrquizo@soton.ac.uk

Abstract. Many natural and engineered structures possess cellular and porous architecture. This paper is focused on the mechanical characterisation of additively manufactured lattice structures. The lattice consists of a stack of polylactic acid (PLA) filaments in a woodpile arrangement fabricated using a fused deposition modelling 3D printer. Some of the most promising applications of this 3D lattice material of this type include scaffolds for tissue engineering and the core for sandwich panels. While there is a significant body of work concerning the manufacture of such lattice materials, attempts to understand their mechanical properties are very limited. This paper brings together manufacturing with the need to understand the structure-property relationship for this class of materials. In order to understand the elastic response of the PLA-based lattice structures obtained from the fused deposition modelling process, single filaments manufactured using the same process were experimentally characterised first. The single PLA filaments were manufactured under different temperatures. These filaments were then characterised by using tensile testing. The stress-strain curves are presented. The variability of the measured results is discussed. The measured properties are then taken as input to a finite element model of the lattice material. This model uses simple one-dimensional elements in conjunction with a novel method achieving computational economy which precludes the use of fine meshes. Using this novel model, the apparent elastic modulus of lattice along the filaments has been obtained and is presented in this paper.

1. Introduction

Many natural structures such as trees and skeletons are made of cellular materials such as bone and wood [1]. Engineers and builders used cellular materials for several centuries such as wood or cork in various applications [1]. However, modern synthetic cellular solids require design, analysis, manufacture and optimisation. An apparent property of this class of materials is the relative density, defined as the density of the cellular solid divided by that of the solid it is made from [2]. When properly architectured, the materials show high stiffness to weight ratio [3].

Additive manufacturing has emerged as an option for fabricating cellular materials, since the fabrication of lattices with customised 3D-shapes has been developed [4, 5]. The manufacture process employed in this work is fused deposition modelling (FDM). FDM is a well-known additive manufacturing (AM) process used to produce prototypes and more recently, end use parts [6]. The parts are built by means of extrusion of filaments through nozzle head. The extruder head moves following a specific path on the printing plane (XY plane) then moving

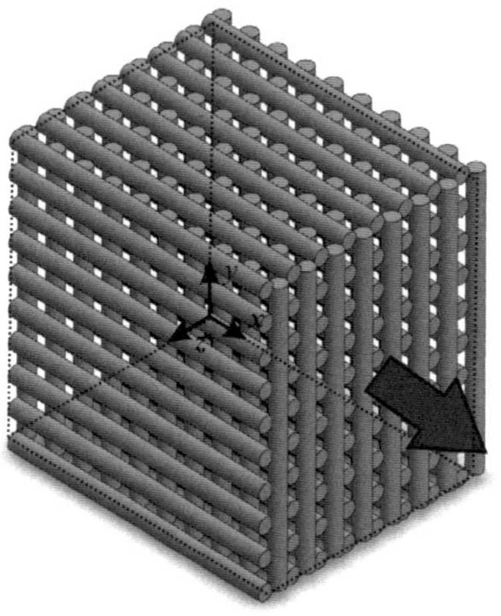

Figure 1. CAD model of the lattice structure.

along the axis perpendicular to the printing plane (Z axis), resulting in a layer by layer manufactured part. Due to the thermal processes taking place during the printing procedure, each layer bonds with the previous layer extruded.

In tissue engineering, scaffolds require optimal biological characteristics to facilitate the proliferation of the cells. They also are required to have mechanical properties sufficient to withstand loading due to the ingrowth of tissue during their use [5, 7]. Initial attempts to understand and predict the mechanical behaviour of this type of lattice structures used as scaffolds for tissue engineering have been presented in [4, 8, 9]. In [4, 8], finite element analysis (FEA) is used to predict the mechanical behaviour of scaffolds fabricated using robocasting and selective laser sintering. The models were meshed using tetrahedral and brick elements respectively. A model to obtain the elastic moduli of the scaffold desired as a function of the diameter of the filaments, spacing and overlap between layers is presented in [9].

The woodpile architecture can also be found in sandwich plate structures used in aerospace where lightweight and specific stiffness are required [10, 11]. In [10, 11], titanium and nickel-based cylinders in woodpile and woven architectures form lattices used as cores in a sandwich plate structures. Experimental tests, compression and shear, have been conducted on titanium composites-based structures [10]. Models to understand the elastic response of woven lattice structures are presented and compared with experimental measurements on stainless steel structures in [11]. From the literature reviewed, it emerges that despite the flurry of activites in the area of additive manufacturing, studies on the mechanics of material thus produces are limited.

The structure of this paper is as follows. In order to characterise the elastic response of the lattice material, single filaments were manufactured via FDM and tested under tensile loads. The measured stress-strain curves are presented in Section 3. The Young's modulus obtained from these curves is then used as input for a novel finite element model to calculate the elastic response of the lattice structure along the filaments direction. The FEA model is presented in Section 4. Finally, conclusions and future work are presented in Section 5.

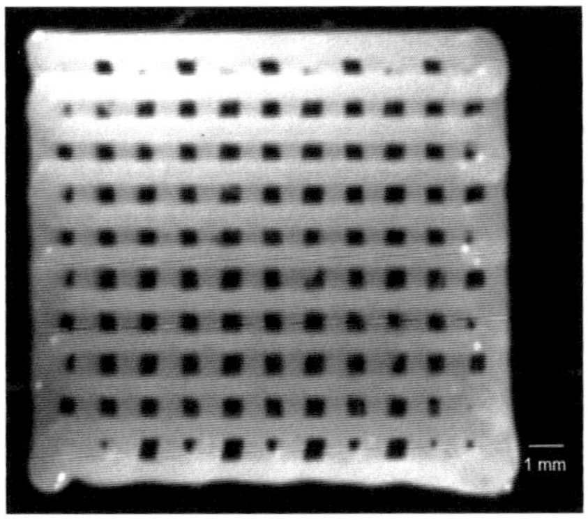

Figure 2. 3D Printed lattice material.

2. Fused deposition modelling as a manufacturing process

As mentioned earlier, the lattice structure of interest within this paper is fabricated using FDM. FDM, as with most other common additive manufacturing techniques, makes use of depositing the material in a semi-molten state. Material in this state is deposited spatially in a manner guided by a set of computer instructions known as G-codes that drive the nozzle. Each layer of the material thus extruded then fuses with the adjacent layers previously deposited. In order to have a complete control on the internal architecture of the material, our approach here is to use a MATLAB code first, which generates the G-codes for the required internal geometry. Using this code it was possible to control the distance between each filament, in the XY plane, the distance between each layer, the number of layers, the number of filaments parallel to the X axis, and parallel to the Y axis. In contrast, standard additive manufacturing techniques rely on the instructions generated by the machine once a CAD-file is provided to the AM machine.

In the past, there have been several studies related to the manufacture of parts using various additive manufacturing processes. The attention of previous authors on porosity control and the properties of components thus manufactured, is limited. Analysis on the mechanical properties of parts manufactured using FDM can be found in [6, 12–17]. However, in these studies, there was no attempt to control the porosity of the manufactured parts.

A final 3D printed lattice is shown in Figure 2. This printed lattice has porosity of 1×1 mm in the printing plane and 10 layers in the axis perpendicular to the printing plane. The same manufacture parameters used to print the lattice were used to print single filaments. These filaments were then submitted to tensile tests. The results are presented in the next section.

3. Single filaments characterisation

The first step towards the characterisation of the lattice structure is to test a single filament having the same cross-section, material and manufacturing conditions (e.g. temperature) as the actual parts. DEBEN MICROTEST ® module was used for tensile testing (see Fig. 3). The testing module consists of two clamps that can be brought close to each other up to 10 mm apart. Further extension up to 10 mm can be achieved by a controlled motor with displacement and force sensing transducers. Using the software module, load-extension curves of the single filaments were obtained. The data were imported into MATLAB for further processing. From the displacement and force information provided by the tensile testing machine, stress and strain were calculated after incorporating cross-sectional information and the gauge length of the samples.

The cross-sectional area has been approximated as an ellipse as it appeared reasonable from

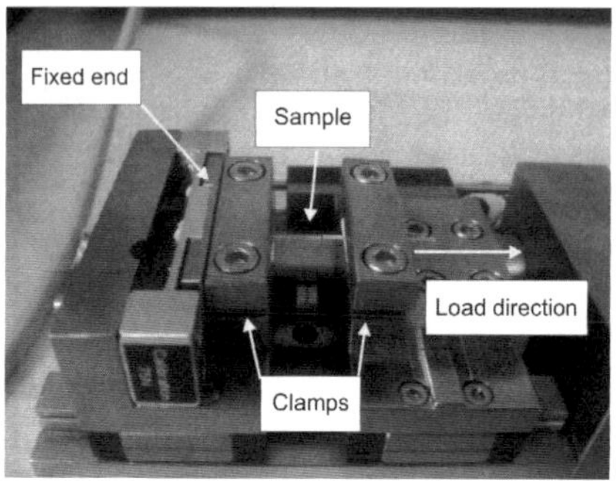

Figure 3. The Deben MICROTEST ® module showing the principal components and a single filament. The sample shows failure away from the two clamped ends.

Figure 4. SEM image of the fractural surface showing adhesion with the adjacent layer.

the appearance (Figure 4). The minor and major axes for 10 filaments were measured using a macroscope. The mean area was obtained as $1.8167 \times 10^{-7} \mathrm{m}^2$. This value for the area was later used to obtain the stress for plotting the stress-strain curves.

Single filaments were tested under quasi-static conditions of slow extension rate. After scaling the force and extension values with respect to the cross-sectional area and the original length, stress-strain curves were obtained and are presented in Figures 5, 6 and 7. These results refer to filaments manufactured at 180 °C, 190 °C, and 200 °C respectively.

From these stress-strain curves, the Young's modulus was obtained. The results are presented in Table 1. The Young's modulus is seen to be fairly consistent for all the cases in the range of $3.0 - 3.5\,\mathrm{GPa}$ (which is within the range of the value reported in the literature [5]). However, the failure stress shows remarkable variability. This variability could be attributed to several factors. One of the most significant sources of this variability appears to be bubble formed during the melting process. This is expected to introduce variability in strength that is very different from sample to sample. Additionally, factors such as variability in the cross-sectional area, the difficulty in their accurate measurement, inevitable variability at the clamping location, etc. have an effect on the variability. Conducting an experimental error analysis, the formula $\sigma = F/\pi r^2$ for the stress calculation, assuming a perfect circular cross-section resulted in: $|d\sigma/\sigma| = |dF/F| + 2|dr/r|$. Note the factor of 2 that appears with the radius term resulting in *twice* the overall error in the stress calculation. If the cross-section is treated as an ellipse instead,

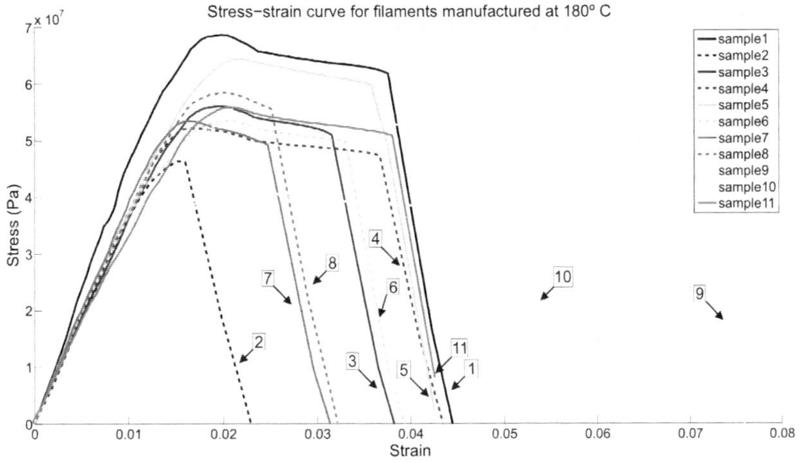

Figure 5. Stress-strain curve for filaments manufactured at 180 °C. For convenience, numerical labels with each curve indicate the sample number.

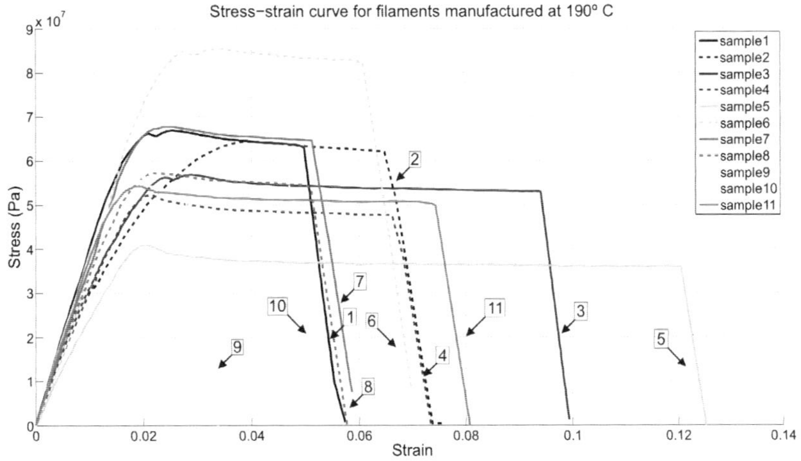

Figure 6. Stress-strain curve for filaments manufactured at 190 °C. For convenience, numerical labels with each curve indicate the sample number.

the formula for stress changes to $\sigma = F/\pi ab$ where a and b are the major and minor radii of the ellipse, we obtain $|d\sigma/\sigma| = |dF/F| + |da/a| + |db/b|$ where the errors in the measurements of the minor and major axis of the ellipse will add up affecting the the final percentage error. This highlights the sensitivity in the measurements of the cross-sectional dimension to the calculated stress. It is envisaged that this may have contributed to the scatter in the modulus of elasticity as observed from different samples. Further study to ascertain this conjecture is under way. The values of the material properties obtained from single-filament tests are subsequently used in the next section for the modelling of the complete lattice.

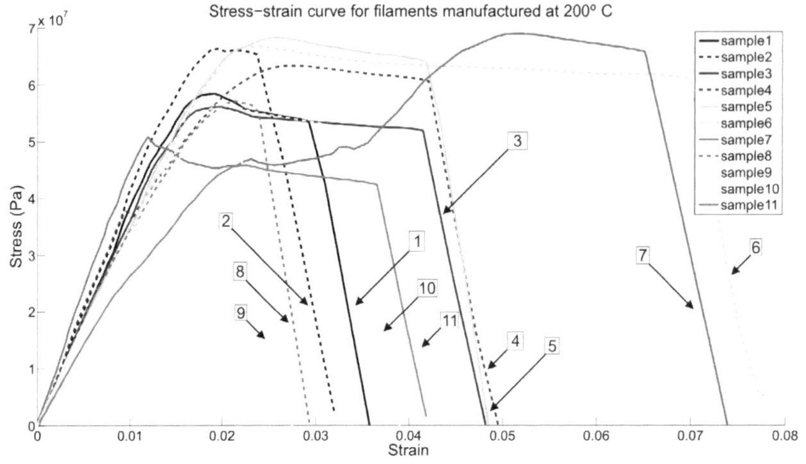

Figure 7. Stress-strain curve for filaments manufactured at 200 °C. For convenience, numerical labels with each curve indicate the sample number.

Table 1. Measured mechanical properties and their variability of single filament samples.

Temperature (deg)	Young's modulus (GPa)	Std dev. (GPa)	Elongation to break (%)	Std dev. (%)	Yield Strength (MPa)	Std dev. (MPa)
180	3.456	0.545	4.28	1.33	55.2	6.94
190	3.215	0.627	7.13	2.37	59.6	10.6
200	3.444	0.592	4.58	1.60	58.1	6.44

4. Micromechanical modelling of the complete lattice

In order to understand the apparent elastic response of the lattice material, when the load is applied in one of the directions parallel to the filaments, a finite element model was created. The model employs one-dimensional elements that have stretch and flexture degrees-of-freedom. This leads to computational economy which precludes the use of fine meshes often used by other authors [4, 8]. While a fine mesh is expected to be more accurate, the present analysis is in favour of simple finite elements as one of the purposes of this work is to develop analytical models for such lattice material. Computationally complex models obscure such understanding.

The method consisted on modelling each segment of the filament using Timoshenko beams (*beam188* in ANSYS). Each segment is a part of a filament with FE nodes placed at the physical joints. Figure 8 presents a portion of the lattice structure. The solid lines represent beams that model individual filaments. The dashed lines are the "virtual" elements proposed to simulate the rigidity between each layer. These are necessary because a beam element is a line in space and the non-intersecting filaments on two adjacent planes need to be kept separated by a fixed distance. This distance is equal to the sum of the radii (r_1 and r_2) of the filament belonging to two such layers. Using 3.25 GPa as the material Young's modulus for the model, and 3000 GPa for the virtual element representing the bonding. Several calculations were carried out using

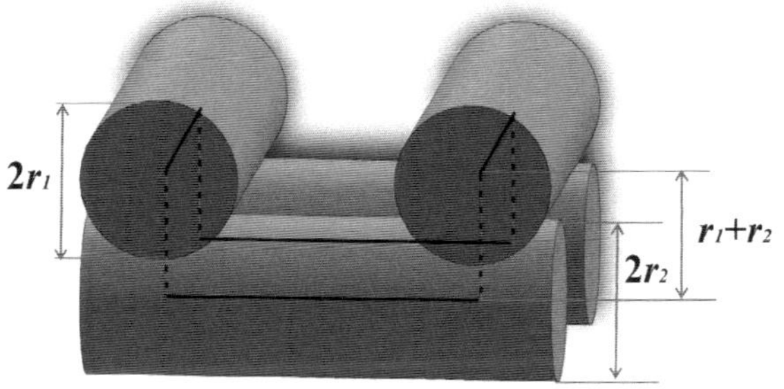

Figure 8. A schematic diagram showing a portion of the lattice structure.

different stiffness values for this virtual element, resulting in negligible differences in the results. This confirms that the virtual elements were stiff enough to be treated as 'rigid'.

The following boundary conditions for the FE simulation were set:

- The nodes lying on the XY plane are constrained to move only in that plane, but not out of it.
- The nodes lying on the YZ plane are constrained to move only in that plane, but no out of it.
- The nodes lying on the XZ plane are constrained to move only in that plane, but no out of it.
- Finally, the origin is constrained in all directions.

The effective elastic modulus along the filaments was obtained as $\bar{E} = 0.596\,\text{GPa}$. This result is validated using the expected effective modulus relation $\bar{E}_e = E_m\varphi$, where φ is the fractional area $\varphi = aN_f/A$. The fractional area is the relationship between the effective area which is the total area of the single filaments cross-sections (a is the area of a single filament and N_f is the total number of filaments normal to the cross-section of the lattice) and the total area of the lattice cross-section (including air gaps). The expected effective modulus is then $\bar{E}_e = 0.5993\,\text{GPa}$, which is clearly in agreement with the simple FEA model used here.

5. Conclusions and future work

Mechanical characterisation of lattice material arranged in a woodpile configuration was carried out. Stress-strain curves were measured in the laboratory to characterise single filaments, first. The variability in the elastic response of the filaments, manufactured under nominally identical conditions and carefully tested, was observed. Several reason for this variability, such as, different samples due to manufacture inconsistency, error in measuring the cross-sectional area, and clamping of delicate samples, were identified. For the present study, a nominal cross-sectional area was used to calculate the stresses. In future analyses each sample cross-section will be used for these calculations. The effect of the variability of the measured cross-sectional area in the apparent mechanical property is currently under way. A stochastic model for such lattice material is also being developed.

The effective moduli of the complete lattice along the struts were obtained using a finite element analysis. This was achieved relatively cheaply and effectively by the use of one-dimensional elements in conjunction with 'virtual element' that supposedly are 'rigid'. However the elastic response along the axis perpendicular to the printing plane is significantly more challenging. Work is in progress in this direction. The bonding between two filaments is likely to play an important role. This aspect of mechanics is currently being studied.

Acknowledgement
This work was supported by the Grants of the Mexican Science and Technology Council (CONACyT) for a PhD degree program at University of Southampton, UK.

References

[1] Gibson L J and Ashby M F 1999 *Cellular solids: structure and properties* (Cambridge university press)
[2] Gibson L J 2005 *Journal of Biomechanics* **38** 377–399
[3] Banerjee S and Bhaskar A 2009 *International Journal of Mechanical Sciences* **51** 598–608
[4] Miranda P, Pajares A, Saiz E, Tomsia A P and Guiberteau F 2008 *Journal of Biomedical Materials Research Part A* **85** 218–227
[5] Yang S, Leong K F, Du Z and Chua C K 2001 *Tissue engineering* **7** 679–689
[6] Bagsik A, Schoppner V and Klemp E 2010 *14th International Scientific Conference on Polymeric Materials*
[7] Hutmacher D W, Sittinger M and Risbud M V 2004 *TRENDS in Biotechnology* **22** 354–362
[8] Cahill S, Lohfeld S and McHugh P 2009 *Journal of Materials Science: Materials in Medicine* **20** 1255–1262
[9] Norato J A and Wagoner Johnson A J 2011 *Journal of biomechanical engineering* **133**
[10] Moongkhamklang P, Deshpande V S and Wadley H N G 2010 *Acta Materialia* **58** 2822–2835
[11] Queheillalt D T, Deshpande V S and Wadley H N G 2007 *Journal of Mechanics of Materials and Structures* **2** 1657–75
[12] Ahn S H, Montero M, Odell D, Roundy S and Wright P K 2002 *Rapid Prototyping Journal* **8** 248–257
[13] Rodríguez J, Thomas J and Renaud J 2001 *Rapid Prototyping Journal* **7** 148–158
[14] Magalhaes L, Volpato N and Luersen M 2014 *Journal of the Brazilian Society of Mechanical Sciences and Engineering* **36** 449–459
[15] Kulkarni P and Dutta D 1999 *Journal of Manufacturing Science and Engineering* **121** 93–103
[16] Li L, QSun, Bellehumeur C and Gu P 2002 *Journal of Manufacturing Processes* **4** 129–141
[17] Bellini A and Geri S 2003 *Rapid Prototyping Journal* **9** 252–264

IOP Conf. Series: Materials Science and Engineering **74** (2015) 012005 doi:10.1088/1757-899X/74/1/012005

Sacrificial component fabrication for optimised production of micro-vascular polymer composite

B Dalton[1,2], D Dixon[1,2,3], A McIlhagger[1,2,3] , E Archer[1,2,3]

[1] School of Engineering, University of Ulster, Shore Road, Newtownabbey, Co. Antrim BT370QB, UK.
[2] ECRE Engineering Composites Research Centre, University of Ulster, Shore Road, Newtownabbey, Co. Antrim BT370QB, UK.
[3] ERI Engineering Research Institute, University of Ulster, Shore Road, Newtownabbey, Co. Antrim BT370QB, UK.

bm.dalton@ulster.ac.uk

Abstract. Smart functional materials are a viable future goal for advanced applications in aerospace, space and medical applications. In this work micro-vascular polymer composite systems have been developed using sacrificial fibres produced from catalyst loaded Poly(lactic acid). The sacrificial fibres have been produced via a published technique which treated PLA in a solvent catalyst mixture of 60% Trifluoroethanol, 40% H_2O dispersed with 10 wt% tin (II) oxalate catalyst. A second process of polymer extrusion of PLA using graded fill contents of tin (II) oxalate has also been developed for the up scaled production of fibres as an alternative to solution treatment. Thermal analysis (TGA) was used to compare sacrificial fibre specimens. PLA fibres produced via the polymer extrusion method outperformed solution treated fibres displaying a lower degradation onset temperature (average 25°C lower), higher degradation rates (observed through a derivative curve comparison) and lower residual catalyst content (0.67% solvent treated fibre against 0.16% extruded fibre). The continuous extrusion process is solvent free and is suitable for high volume production. This work has been carried out to fully understand the fabrication issues with sacrificial components.

1. Introduction

"Smart materials" is a term now used to describe materials with an additional functionality which are able to react to their environment. With composites the dominant areas are self-healing, thermal control, sensory/damage detection, and electrical/magnetic utilization. The definition of a smart material is "one in which a key material property is altered in a controlled fashion in response to the introduction of pre-determined external stimulus" [1]. Self-healing materials were first researched at in the 1990s. Much of the work since has been heavily dominated by R&D, with industrial applications relatively confined. Much of the inspiration for early works was influenced by organic biological systems. This inspiration still dominates the ideology behind smart autonomous materials.

Traditional methods of repairing polymer composites are both basic and time consuming. Added to the repair process, the input of inspection and detection and the process becomes a costly, non-value added affair. Techniques of welding, patching and in-situ curing are all employed currently. These methods are useful, and may never be fully superseded [2]. While intelligent polymer composite materials have been viewed by many industries, major interest has come from the aerospace sector.

Many of the features promised by smart materials can apply directly to aircraft applications. These may include structurally integrated batteries for energy storage, structural profile morphology for aerodynamic application in wing skins, electrical and magnetic surface signatures for signal and transmitting, integrated sensors for detection and monitoring, thermal control and, of course, self-healing capabilities [3].

There are limited techniques for achieving vascular networks for functional fluid delivery. Hollow glass fibre tubes which are pre-impregnated with epoxy have been used. Known as discrete channel arrays, the glass pipettes were incorporated into glass and carbon fibre polymer composites with limited success. After mechanical testing to induce 'healing', samples with completed heal cycles showed no more mechanical recovery than control samples with unfilled fibres [4]. Micro-machining and etching are achievable via various methods also but again with limited success and lengthy integrated process. The latest and most promising technique for creating a vascular network within a composite is that of the incorporation and subsequent thermal removal of sacrificial fibres.

2. Methods and materials

2.1. List of materials

PLA monofilament of 500 μm diameter was used as received from Monofil Technik UK. The catalyst tin (II) oxalate and solvent Trifluoroethanol were used as received from Sigma-Aldrich UK. The epxoxy Huntsman epoxy system; Araldite LY 564 with Aradure 2954 cure agent was used for both carbon fibre and epoxy optical samples.

2.2. Tools and analysis

TGA was performed on a TA Instruments Q100 Series machine. For Scanning Electron Microscopy, an FEI Quanta 200 was employed. An Olympus SZ-PT was used for optical imaging of incorporated PLA fibres and post vaporisation evacuated chambers. Vascular fluid behaviour was also studied via optical microscope for imaging and video footage. Carbon fibre composite components were produced on a 500 mm^2 (bottom) aluminium tool piece, with 390 mm^2 caul plate, sealed with Nylon 800G vacuum bagging material from Aerovac systems, giving composite components averaging 350 mm^2. Both vascular (study) and non-vascular (reference) composite components were consolidated via Resin Transfer Moulding (RTM). An Edwards RTM system was employed for the transfer moulding configured with a Walther Pilot degassing system type MD6 12 HZM. Vaporisation of sacrificial components VaSc (post consolidation) was carried out via two methods. A vacuum bagging technique was employed as well as a sealed vessel vacuum. 800G Nylon vacuum bagging was used (suited for larger more complex geometrical components). Also a Pyrex 10 lt vacuum vessel was used for smaller components. A Magnum Venus Plastech 240 v twin port vacuum system was used for VaSC experiments. For alternative fibre production methods, (polymer extrusion with catalyst filler) a Thermo Scientific Haake Polylab OS RheoDrive 7 (drive system) with Haake RheoMex OS ptw16 (barrel) was used.

2.3. Fibre treatment (solvent)

The fibre treatment method previously used to prepare vascular composites is not fully described in literature [5]. For the initial experiments 500 μm PLA monofilament was rolled into loose coils and supported within a 3 litre Pyrex beaker on a steel wire nest. The beaker, wire nest, flea and thermometer were all solvent cleaned to minimise contamination in the catalyst solution. The beaker was placed onto a heater plate/stirrer [Figure 1]. A mixture containing 60% trifluoroethanol, (TFE) and 40% H_2O dispersed with 10wt % tin (11) oxalate was mechanically stirred for 30 minutes to evenly disperse the catalyst. All catalyst mixtures were measured via weight on a digital scale and the solution was then poured into the beaker fully submerging the suspended monofilament. The hot plate was set to 37°C and regulated via a thermostat. To keep the catalyst suspended during the 24 hour soak a medium speed stir setting was selected. All fibre treatment was performed in a fume cabinet and with appropriate personal protective equipment due to the corrosive properties of trifluoroethanol. After a 24 hour soak time at 37°C the fibre bundles were removed from the solution and suspended within the fume cabinet to air dry for a further 24 hours. Evaporation of the trifluoroethanol/catalyst mixture (discussed in results section) was minimised in by inclosing the mixture within a sealed vessel.

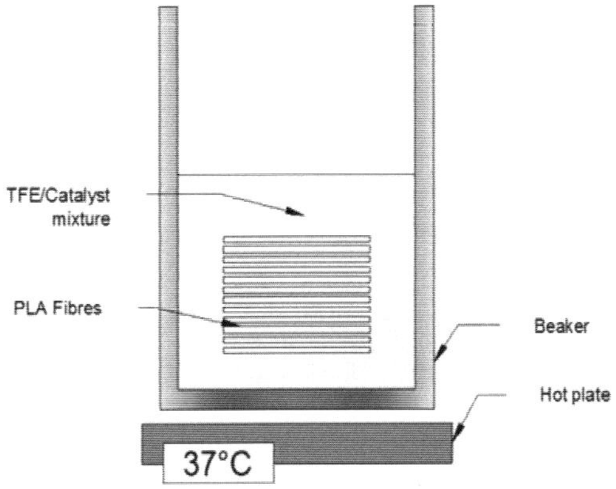

Figure 1. Solution treatment of PLA fibres on hot plate.

2.4. *New fibre production method*
For primary experiments to determine the polymer melt/catalyst compatibility, PLA material (in fibre form) was re-melted in a glass beaker at 100°C. Tin (II) oxalate was added to the polymer melt and dispersed roughly by stirring. Several oven runs were necessary due to the rapid cooling of the PLA when removed from heat. Crude fibre like structures could then be hand drawn into various lengths and diameters producing structures which could be tested for vaporisation efficiency. PLA/catalyst structures were suspended in a matrix material of Huntsman epoxy (Araldite LY 564 with Aradure 2954 hardener) and room cured for 24 hours before a post cure oven cycle of 2 hours at 120°C. The resulting sacrificial structures could then be trimmed to expose fibre ends and put through a fibre removal cycle. Prepared components were sealed in a Pyrex vacuum vessel and run at 180°C for a 20 hour oven cycle under vacuum. The process left a vascular epoxy sample with visible cleared channel architecture.

Following this, fibres were produced via a polymer extrusion method using Tin (II) oxalate (a powder) as a filler/additive in the extrusion process on a Thermo Scientific Hakke PolyLab OS RheoDrive 7 (drive system) with Rheomex OS PTW16 barrel. The extrusion of a thermoplastic monofilament [6] becomes more temperamental when incorporating a powder filler [7,8]. The same monofilament from solvent treatment was used as a polymer feed. Using the same PLA material as a base would allow for a closer comparison of the two end products. PLA fibre reels were mounted on spindle mounts and six fibre ends were fed into the barrel. A slow screw speed ensured the fibres would be taken into the barrel and catalyst material was added using a screw-driven force feeder with 7 different feed rates ranging from <0.5% to 5.0%. A water tank cooled the fibre upon exiting the die.

With these extruded fibres, vascular components were developed in two forms; carbon fibre reinforced vascular structures, and plain epoxy vascular. Carbon fibre reinforced samples were developed to determine the mechanical implications of vasculature on a load bearing structure, and epoxy samples were developed for optical analysis, to view the void architecture. Vaporisation of sacrificial components (VaSC) techniques follow the reported literature method [4]. Embedded sacrificial components were thermally evacuated through an oven cycle of 20 hours at 180°C under vacuum.

3. Results and discussion

3.1. *Fibre treatment (solvent)*

Untreated PLA monofilament is a rigid transparent material (Figure 2a), whereas solvent treated fibres turned white and were more ductile (Figure 2c). Long exposure times to the PLA solvent trifluoroethanol swells the fibre allowing the tin catalyst to absorb but over exposure (treatment times above 24 hours or at temperatures higher than 37°C) dissolved the PLA fibres into a white paste like material. As an initial indicator a uniformed white fibre may be a sign of a successful treatment but upon closer inspection fibres were not uniformly discoloured throughout their entire lengths. Observable clear patches (Figure 2b) would suggest areas where catalyst treatment has not fully occurred.

3.2. *Fibre extrusion*

The first identified issue with extruded fibres was a significantly more brittle material when compared to solution treated fibres. This is a result of two factors. Firstly the thermal and mechanical (shear) action experienced through the extrusion process where the molten poly(lactic acid) is subjected to barrel temperatures of >200°C. The result of these events contribute to a loss in mechanical properties for the extruded fibre. Secondly the catalyst powder fill content has a direct effect on the polymer during cooling [7, 8]. This can be confirmed by the fact that higher fill content fibres are more brittle than low fill content fibres. From results there is an optimum fill content (0.5% - 1.0%) after which, higher levels of catalyst become a hindrance in the evacuation of chambers (Figure 3) resulting in lower percentage clearance. High levels of residual catalyst materials contribute to 'plugging' and can reduce the fibre removal efficiency. These results correlate with findings from TGA and SEM, where there is a direct relation between fill content and residual materials. However there were different methods employed for the calculation of channel clearance. For solvent treated fibre components, the percentage channel clearance was calculated from the amount of channels which appeared to be fully cleared. The problem with this method is that some channels may have been 95% clear, with a 5% plug, which will render the entire channel inoperative; as this will affect the overall fibre removal efficiency percentage extruded components were quantified via optical microscope analysis of clear epoxy samples where the entire channel could be viewed. Total areal percentages could then be ascertained giving a more accurate value. 0.5% and 1.0% fill rate fibres perform as well as solution treated fibres and so must be forwarded as the most ideal candidates for further development.

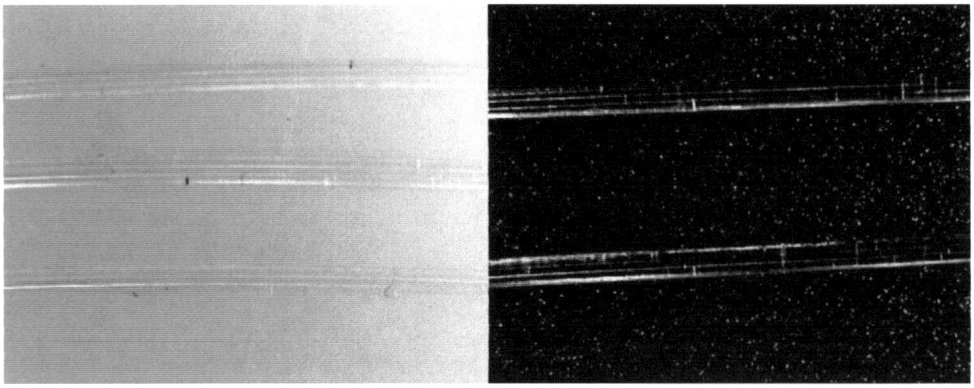

Fig 2. (a) Untreated PLA fibres appearing clear and transparent.

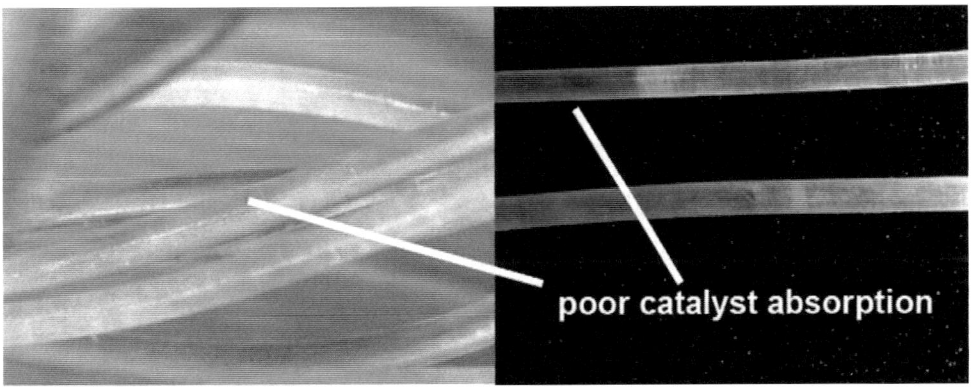

Fig 2. (b) Treated PLA fibres showing poor catalyst dispersion. Poor treatment seen as dark areas on fibres.

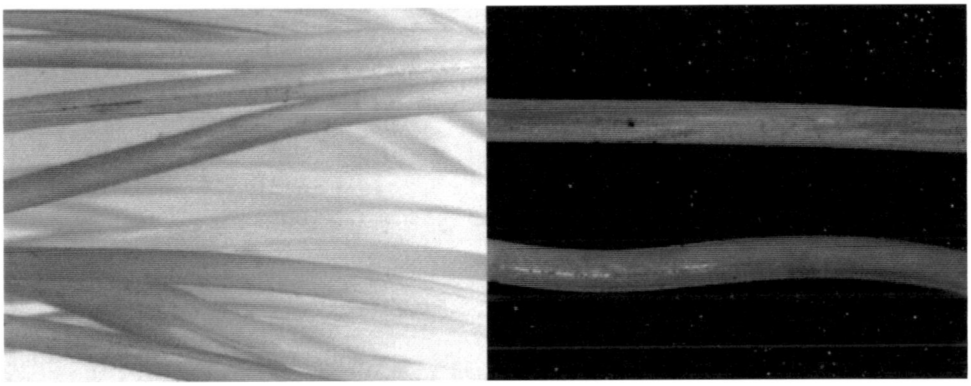

Fig 2. (c) Treated PLA fibres showing good catalyst dispersion.
Successful treatment produced uniformed white fibres.

3.3. *Vaporisation of sacrificial components*
Sample coupons of both carbon fibre and plain epoxy samples were initially developed with 10 micro-channels per sample. Each sample was prepared using the same process and fibres removed via the same technique. Vascular samples were then individually tested for channel content and fluids were driven through each channel; if the channel was clear fluids could travel through the component. If there were blockages the fluid would not travel. The percentage of cleared channels was then recorded (Figure 3). Much higher efficiency for fibre removal was observed for the epoxy only samples versus the samples containing carbon fibre/ epoyy. The vaporisation capability for the sacrificial fibres must be considered equal as all fibres came from the same treatment process and components also experienced identical fibre removal conditions. This would suggest the presence of carbon fibre materials may in some way hinder the evacuation of chambers. Another possibility is that carbon fibre samples have increased 'plugging' due to increased crimping of sacrificial fibres in the presence of carbon.

Cross sections of vascular carbon fibre samples were observed in Figure 4b. Whilst structural patterns were not perfect there has been a good average areal displacement of the channels. Also noted is the increased crimp in vascular samples (Figure 4b) when compared to non-vascular (Figure 4c) due to the presence of the carbon fibres. Sacrificial components displace carbon fibres in the layup. Upon removal of sacrificial fibres the carbon fibres remain displaced. It is not yet known if this crimping has an effect on the mechanical properties of the component. Optical/epoxy samples were developed to observe the internal void architecture of vascular components. Channels were highly visible allowing for the mapping of 'plugs' and rough estimates of fibre removal efficiency. Channel architectures including a double braid (Figure 5a) triple braid (Figure 5b) and quadruple braid (Figure 5c) have been realized.

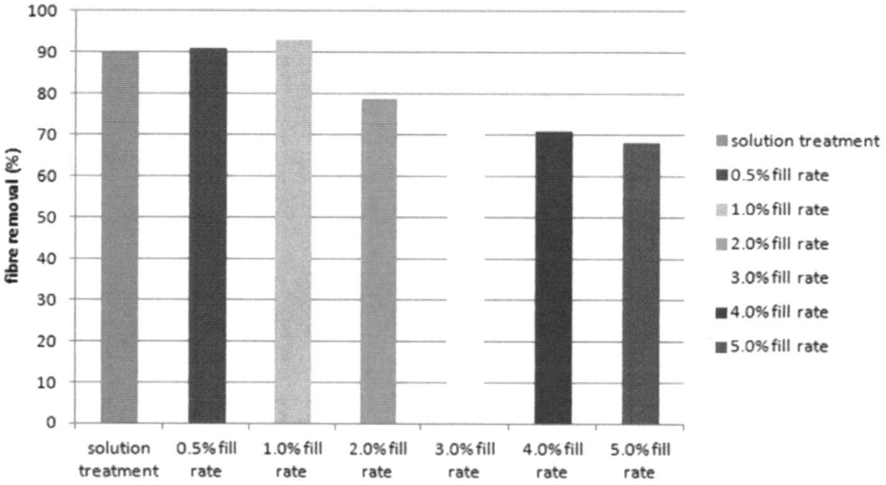

Figure 3. Fibre removal (%) for extruded components of varying fill content.

4. Thermogravimetric analysis

Thermogravimetric analysis was performed on poly(lactic acid) PLA materials. The samples were prepared to compare PLA fibres developed through the solvent treatment method against those prepared through polymer extrusion with catalyst filler. The thermal degradation characteristics of treated PLA materials will have a serious implication on the efficiency of fibre removal as the most important event of thermal degradation of solvent treated fibres from literature is the onset of mass loss [5].

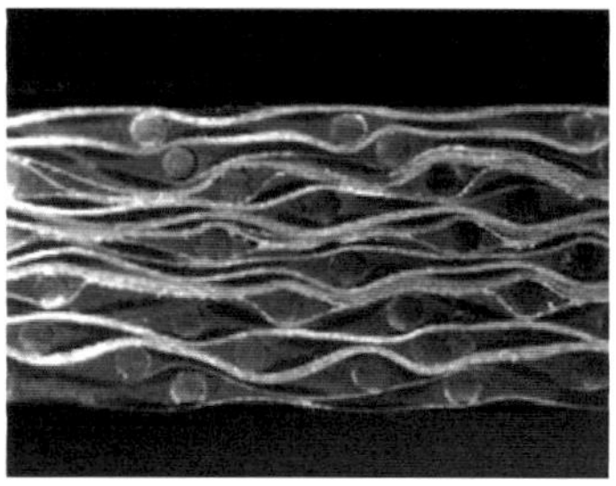

Figure 4. (a) Carbon fibre cross section pre VaSC showing PLA fibre ends.

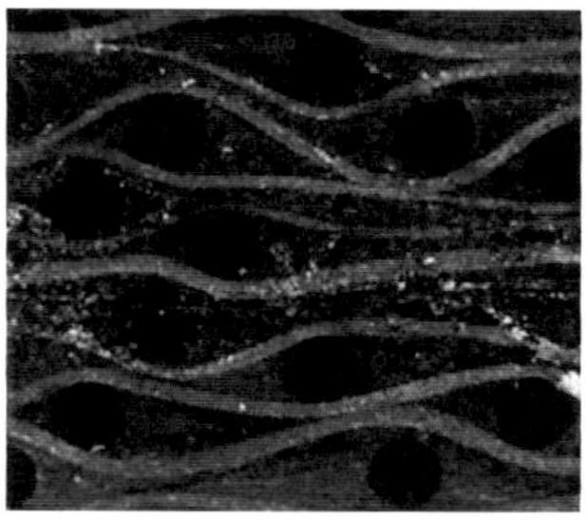

Figure 4. (b) Vascular sample showing displacement of carbon.

Figure 4. (c) Virgin sample with no fibre displacement.

The lowest reported onset of thermal degradation from literature for PLA has been 180°C. The lowest degradation onset temperature achieved with extruded materials in this study was 200°C (Figure 6) with maximum onset of ~300°C. For all runs untreated PLA gave similar results and are comparable to literature. Extruded samples with higher fill content (>1.0%) degraded at lower temperatures however, with a large remaining volume of catalyst filler (tin II oxalate) which degrades at much higher temperatures. The remaining measured catalyst materials relate directly to the percentage fill rates used during production. Also noted from both literature graphs and experimental graphs is the initial weight drop off at ~100°C. This drop is exclusive to solvent treated materials and not observed for extruded materials (seen in all graphs; green curves). It is presumed that this drop off may be the vaporisation/escape of trapped trifluoroethanol (TFE) within the PLA. TFE has a boiling point of 74°C, however trapped materials may take longer to evacuate.

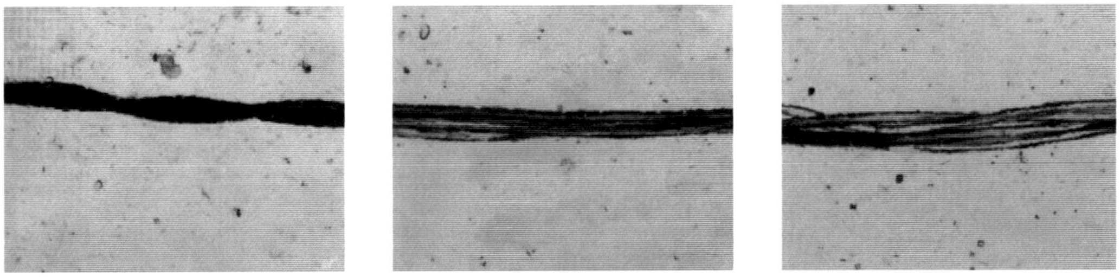

Figure 5. (a) Epoxy specimen with double braid configuration. **Figure 5.** (b) Epoxy specimen with triple braid configuration. **Figure 5.** (c) Epoxy specimen with quadruple braid configuration.

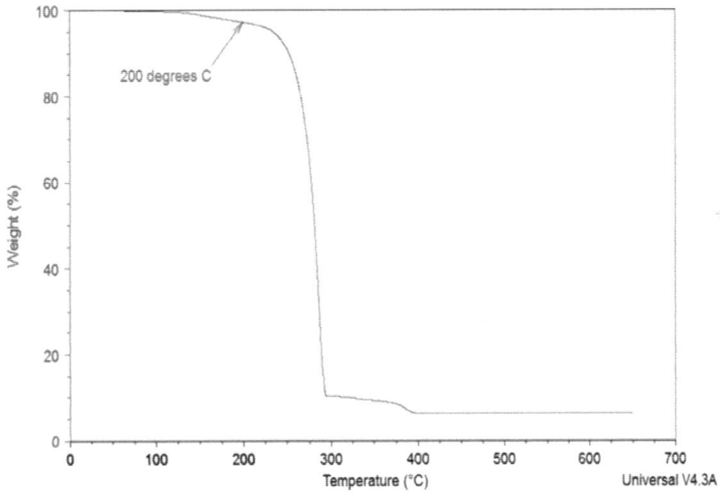

Figure 6. Mass loss onset for extruded PLA with catalyst fill.

4.1. *Degradation profile*

After preliminary runs secondary runs were performed on specimens with the most potential for further trials (0.5% fill rate extrusions); multiple runs were completed on 0.5% samples and solvent treated samples [fig.8]; 1%, 3% and 5% specimens were also run for reference. Mass loss onset for some specimens remain similar in run 2, however, 0.5% fill rate samples experience an earlier onset of mass loss though rates of degradation remain similar for both runs. The remaining specimens with higher fill content (3%, 5%) experience earlier mass loss onset but with significantly larger volumes of residual catalyst materials. Specimens with 1% fill rate display attractive qualities of early onset and low residual catalyst content.

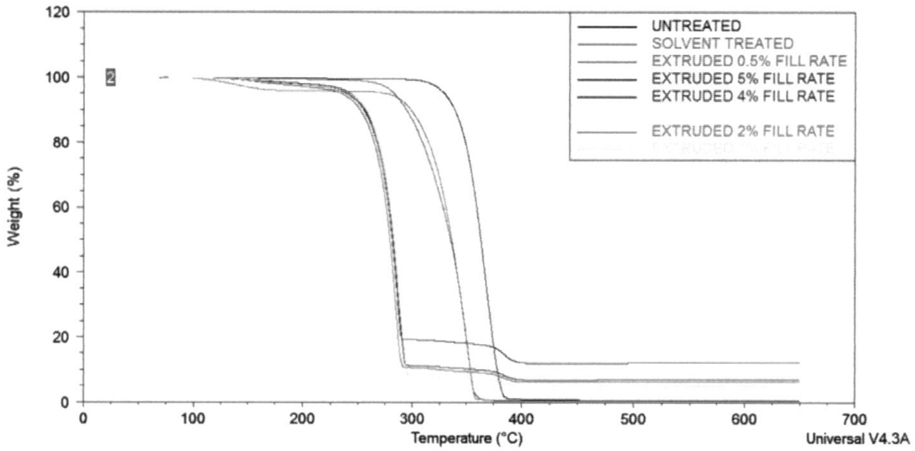

Figure 7. TGA overlay comparison for untreated, solvent treated and extruded PLA materials.

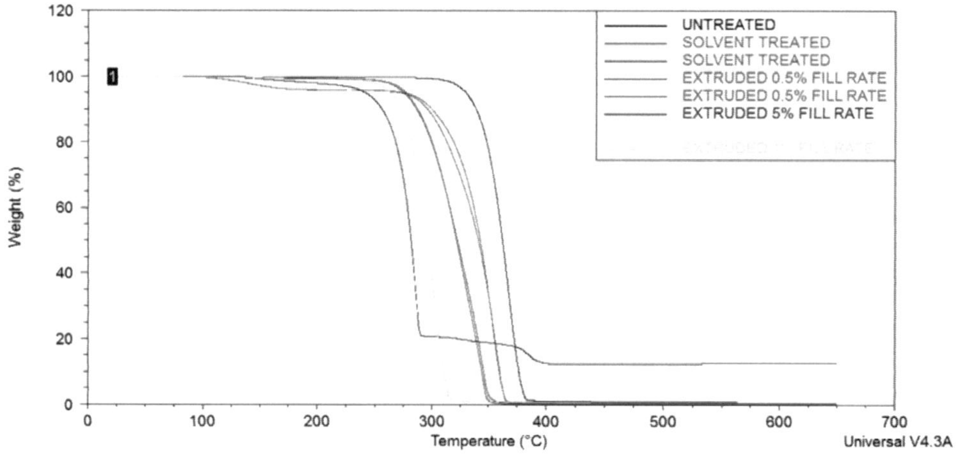

Figure 8. TGA run focusing on most potential specimens of 0.5% extruded against solvent treated.

4.2. *Residual materials*
Whilst 1% fill rate specimens display improved onset temperatures when compared to 0.5% fill rate samples there is significantly less residual catalyst in 0.5% samples (0.16%) when compared to both 1% fill rate samples (0.67%) and solvent treated samples (0.67%) (Figure 9). This data corroborates scanning electron microscopy SEM images taken from interior channel walls of vascular components created using fibres from both solvent treatment methods and extrusion methods.

4.3. *Derivative*
In Figures 10, a sharp derivative peaks represent rapid rates of degradation. 0.5% fill rate samples were compared to 5% fill rate samples (lowest, highest) to determine the effect of catalyst content on degradation rates. It is clear from results that are direct relations between fill content and degradation rates. Higher fill rates (5%) (Figure 10b) degrade faster than low fill (0.5%) (Figure 10a). The

difference is clear in the overlay (Figure 10c). Although faster degradation may be an attractive quality, the higher levels of residue in high fill samples is much more of a concern.

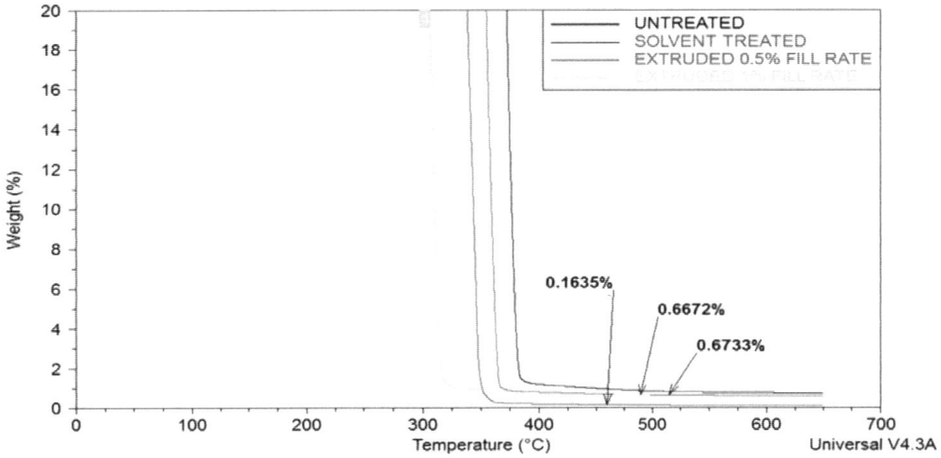

Figure 9. Residual mass readings from TGA.

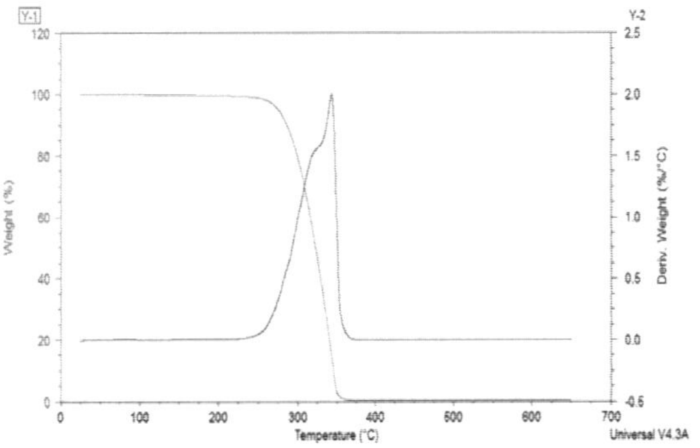

Figure 10. (a) Low fill specimen derivative.

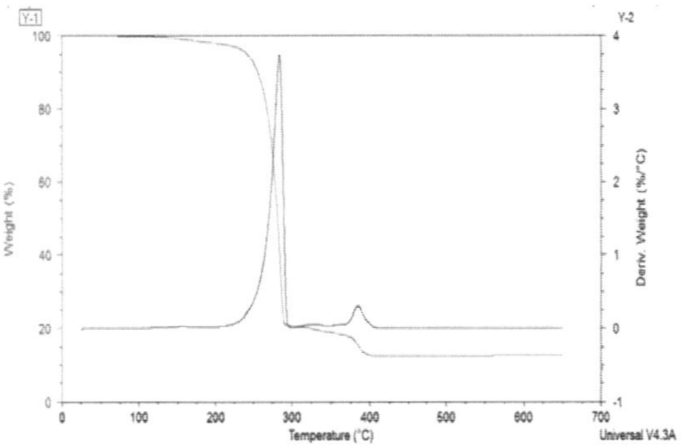

Figure 10. (b) High fill specimen derivative.

5. Conclusions

"Smart functional" polymer composites have been realised. In this study Micro-vascular components have been developed from fibres produced via a solution treatment method previously described in literature, and a novel method of polymer extrusion using a catalyst filler. Solvent treatment of poly(lactic) produced soft and ductile fibres making them ideal for weaving/handling, however, the treatment process is highly critical to the outcome of the sacrificial component. Treatment times are long (24 hours) and solution mixture/temperature must be kept constant (TFE and water has different rates of evaporation). Variability in the catalyst adhesion was found to be an issue which could cause poorly treated areas, ultimately resulting in plugging and channel clearing issues. It is difficult to determine the final catalyst percentage or indeed control catalyst content through solution treatment methods; polymer extrusion on the other hand is a well established versatile process. Components may be produced in a variety of geometrical profiles or indeed injection moulded into more advanced sacrificial components and fibres may be produced with varying levels of catalyst fill. Moreover, if this fill content is optimised, components can be produced yielding less residual materials post evacuation (0.16% for 5% samples against 0.67% for solution treated fibres), have higher channel clearing efficiency (observed through TGA) and cleaner internal channel walls (observed through SEM) than those developed through solution treatment style).

TGA degradation profiles show extruded components with earlier onset temperatures, higher rates of degradation and lower residual catalyst content post degradation. The most promising specimens being the lower fill content materials (0.5% - 1.0%).

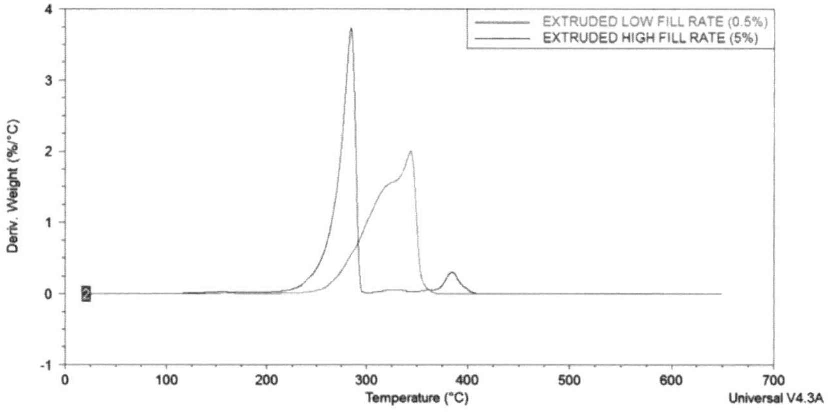

Figure 10. (c) Comparative derivative signal overlay for low fill against high fill specimens.

References

[1] Murphy EB, Wudl F, 2010, "The world of smart healable materials." *Progress in Polymer Science* **35** 223–251.

[2] Wu DY, Meure S, Solomon D, 2008, "Self-healing polymeric materials: A review of recent developments." *Prog. Polym. Sci.* **33** 479–522.

[3] Gibson RF, 2010, "A review of recent research on mechanics of multifunctional composite materials and structures." *Composite Structures* **92** 2793–2810.

[4] Olugebefola SC, Arago NAM, Hansen CJ, Hamilton AR, Kozola BD, Wu W, Geubelle PH, Lewis JA, Sottos NR, White SR, 2010, "Polymer Microvascular Network Composites." *Journal of composite materials* **44** 2596-97.

[5] Dong H, Esser-Kahn AP, Thakre PR, Patrick JP, Sottos NR, White SR, Moore JS, 2012, "Chemical Treatment of Poly(lactic acid) Fibers to Enhance the Rate of Thermal Depolymerization." *ACS Appl. Mater. Interfaces* **4** 503–509.

[6] Spiridonov P, Lambrinos E, Peng Z, 2005, "Extrusion of monofilaments of thermoplastic elastomers." *Synthetic Metals* **152** 61-64.

[7] Bleach NC, Nazhat SN, Tanner KE, Kellomaki M, Tormala P, 2002, "Effect of filler content on mechanical and dynamic mechanical properties of particulate biphasic calcium phosphate-polylactide composites." *Biomaterials* **23** 1579-85.

[8] Liua X, Khorb S, Petinakisb E, Yua L, Simonc G, Deanb K, Bateman S, 2010, "Effects of hydrophilic fillers on the thermal degradation of poly(lactic acid)" *Thermochimica Acta* **509** 147-151.

IOP Conf. Series: Materials Science and Engineering **74** (2015) 012006 doi:10.1088/1757-899X/74/1/012006

Service and maintenance damage assessment of composite structures using various modes of infrared thermography

SA Grammatikos[1], EZ Kordatos[2], TE Matikas[3] and AS Paipetis[3]

[1] Department of Architecture and Civil Engineering, University of Bath, Bath BA2 7AY, United Kingdom.
[2] Materials and Engineering Research Institute, Sheffield Hallam University, Sheffield, S1 1WB, United Kingdom.
[3] Department of Materials Engineering, University of Ioannina, 45110 Ioannina, Greece.

paipetis@cc.uoi.gr

Abstract. Carbon Fibre Reinforced Polymers (CFRPs) have been widely used recently in aerospace technology as primary materials. During aircraft operation the combination of thermo-mechanical loads, gradually degrades the material properties. Therefore, the development of reliable and cost effective damage inspection protocols throughout the service life of these structures is of primary importance for their safe function. Within the scope of this study, Infrared Thermography (IrT) was employed with the aim of assessing the structural integrity of composite aero-structures in both maintenance (off-line) and service (on-line) conditions. In the maintenance concept, IrT was employed in Lock-in mode to evaluate artificially induced damage in bonded repaired materials. For this purpose, various configurations of damage were investigated. In the service concept, IrT was employed for the continuous monitoring of loaded repaired structures in order to assess in real-time any progressive evolution of de-bonding which could lead to failure of the bonded patch repair. The experimental results provided evidence that IrT is capable of qualitative and quantitative field assessment of aero-structures.

1. Introduction

Fibrous reinforced composites are being widely used the last decades in various applications in aerospace. Back to the 19th century the first composite materials were developed by wood and natural resins. Nowadays, composites are used more and more as materials for primary structures in demanding applications. The Boeing 787 Dreamliner and the Airbus A350 XWB are examples of extensive use of Carbon Fibre Reinforced Polymers (CFRPs) and Glass Fibre Reinforced Polymers (GFRPs). With respect to aircraft maintenance, the replacement of a structurally degraded part is not always desirable due to the increased cost. Therefore, efficient repair technologies are always in demand for qualification as standard practices in the aircraft industry. Bonded composite repair is a relatively new repair concept primarily proposed by Baker in Australia [1], which proposes the use of composite materials directly bonded onto the cracked surface. Bonded composite repairs offer significant advantages over the conventional mechanical fastened repair methodologies [2]. At the same time, the development of new materials in high performance applications requires effective methodologies for structural integrity assessment during service, operation and maintenance. Non-

Destructive Evaluation (NDE) techniques allow for the assessment of the internal structural state, i.e. defects, anomalies, critical and subcritical damage incidents. A variety of NDE techniques are employed in the aircraft industry. Ultrasonics [3], radiography [3], thermography [4] and Lamb wave [5] methods are the NDE techniques commonly employed in aero-structures when high detection resolution and accuracy is needed. Infrared thermography (IrT) is a full-field and non-contact structural integrity methodology which exploits the infrared thermal radiation emanated by the investigated material surface. IrT is applied using two fundamental modes which refer to the type of excitation, *i.e.* active and passive. In passive mode, no external thermal source is required whereas in active mode, thermal excitation is achieved via an external stimulus. Various stimulation methodologies have been proposed [2]. Optical (lamps) [6], ultrasound [7], eddy current [8], cyclic loading [9] and current injection [4] are the most frequently used types of thermal stimulation. With respect to maintenance, IrT can be easily employed to detect induced damage. In typical composite structures damage is manifested as delaminations, voids, de-bonding etc. While in service, IrT is capable of recording the initiation and propagation of any induced damage in real time. In the case of periodic loading, the mechanical loading waveform is often synchronized with the thermal sensor acquisition frequency. This system allows for the monitoring of the generated internal stresses which ultimately lead to failure.

In this work, IrT was employed to assess damage both in maintenance (off-line) and service (real time) modes. With respect to maintenance, CFRP patches were applied to repair both composite and Aluminum (Al) substrates. In both substrate scenarios, PTFE tapes were introduced in various locations of the repaired structure to simulate delamination, de-bonding as well as poor bonding. Teflon tapes of constant thickness, but various geometries were situated in-between the layers of the repair and in the patch /substrate interface. The repaired coupons were examined with optical IrT. More specifically, optical Lock-in thermography (LT) was adopted to pin-point the damage concepts. In optical LT, the thermal sensor is continuously synchronized with the source of heating, which in this case is an array of optical lamps. With regards to service, an artificially cracked aluminum helicopter wing (the stabilizer of the SW-4, PZL-Swidnik /AgustaWestland) was repaired using a CFRP bonded patch. In this concept, LT was employed on-line during cyclic mechanical loading to monitor the durability and efficiency of the patch repair. In all stages, IrT was capable of detecting artificially induced damage and monitoring damage evolution in real conditions

2. Experimental

2.1. Maintenance assessment; bonded composite repaired Al and composite substrates

Al and composite plates were repaired with CFRP patches. These consisted of, *(i)* 6 aluminum substrate coupons, and *(ii)* 6 composite substrate coupons repaired with 5-ply CFRP patches. For the patch manufacturing, a satin weave fabric provided by (5H SATIN 43280) HEXCEL was employed for patch reinforcement and Epocast 52 A/B for the patch matrix. The orientation of the patch reinforcement was [0/90°/0/90°/0]. The thickness of the Al substrates was 1.6mm. Regarding the composite substrate case, the reinforcement was provided by a pre-preg (MTM®56-cure cycle: 30 min at 120 0C) woven carbon fabric (199 g/m2) by ACG (UK). The thickness was approximately 1.6mm (identical to the Al case).

2.2. Service assessment; bonded composite repair of al helicopter wing

The structure of the stabilizer consisted of an internal core HexWeb CR III-3,8 5/32-10 (5052). The honeycomb thickness diverges from 32mm to 34.2mm. The aluminum skin was manufactured by the 2024 T3 alloy sheet. The thickness of the external surface varied from 0.4mm to 1.5mm. Figure 1 depicts the prepared for test stabilizer as well as a scheme with the dimensions of the wing structure.

After previous analysis and dynamic testing on the stabilizer, a crack was noticed that has started in the vicinity of the one of four adjusting (on the fuselage) bolted joints. To that respect, this area was repaired with a CFRP patch.

In Figure 1b and 1c one can distinguish the applied bonded composite patch. The artificial crack is shown in Figure 1a. As can be seen, the crack was created close to the bolted joint. Figure 1d displays a schematic representation of the wing. For the requirements of the composite patch, the Hexcel 43280s series carbon fabric was adopted. As for the matrix, the Epocast 52 A/B was utilized.

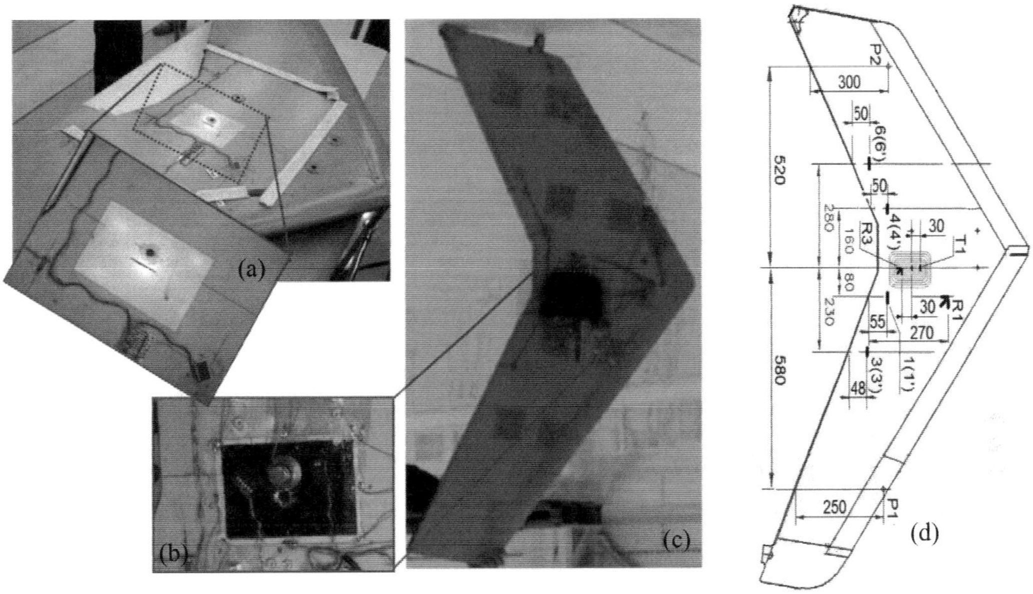

Figure 1. (a) Artificial crack (35mm), patched region, (b) applied bonded patch ($80\times120mm^2$), (c) SW 4 helicopter vertical stabilizer, (d) schematic representation of the stabilizer.

The contained data belongs to PZL-Swidnik S.A. and any further reprinting and usage cannot be made without PZL-Swidnik S.A. written permission.

3. Results and discussion

3.1. Maintenance assessment

Figures 2a *(i-iii)* and 2b *(iv-vi)* depict the thermographic images derived when scanning the repaired plates with LT. Grey and yellow coloured shapes on the coupon configuration schematics represent the location and shape of the artificially induced damages. The thermographic experimental parameters are listed in Table 1.

Table 1. Experimental parameters of thermographic evaluation.

Method	Lock-in Thermography					
Configuration	*i*	*ii*	*iii*	*iv*	*v*	*vi*
Frequency /Hz	0.01	0.1	0.01	0.05	0.05	0.01
Voltage /V	3	2	3	2	2	3
Distance /m	1.18	1.18	1.18	1.18	1.18	1.18
Periods	2	5	2	3	4	2

IOP Conf. Series: Materials Science and Engineering **74** (2015) 012006 doi:10.1088/1757-899X/74/1/012006

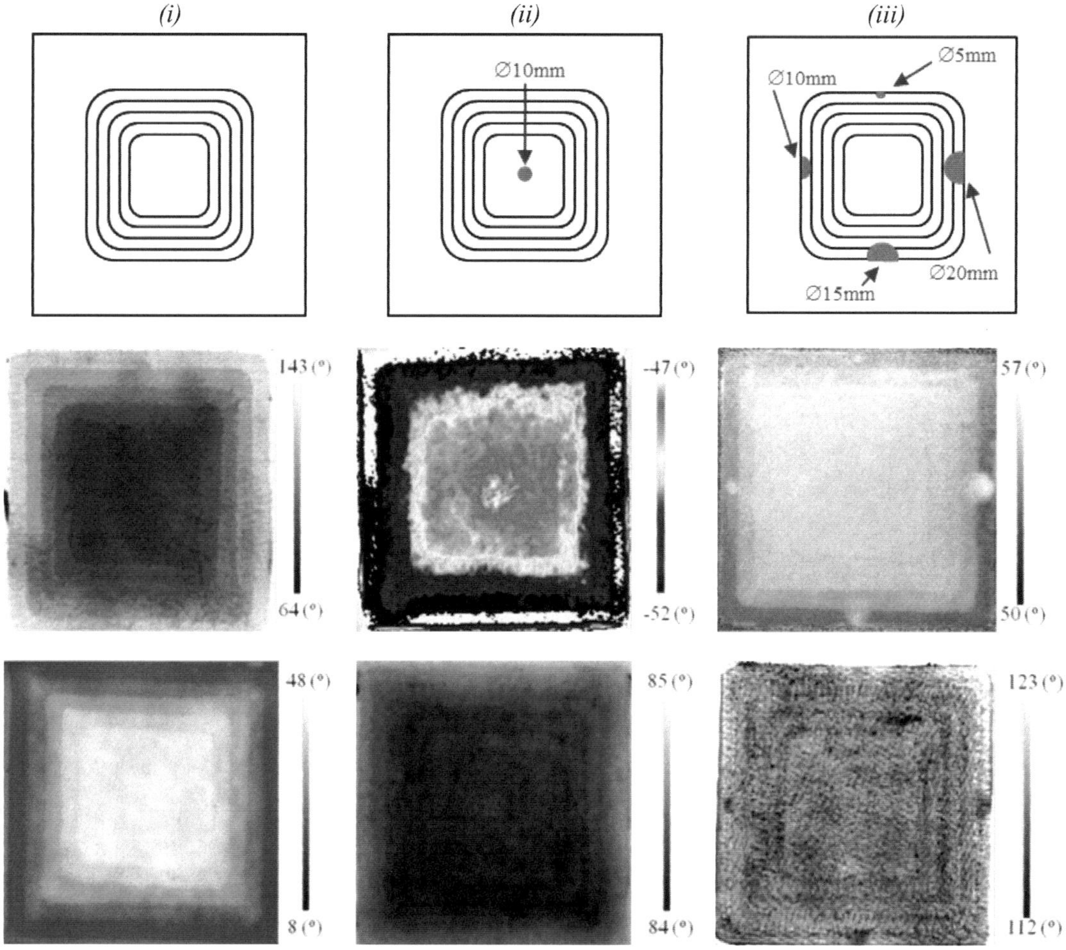

Figure 2a. (A) Coupon configuration, (B) Phase images by lock-in thermography; Al substrate, (C) Phase images by lock-in thermography; Composite substrate, *(i)* No defect, *(ii)* Central patch substrate de-bonding, *(iii)* Patch substrate de-bonding.

As is obvious from Figures 2a and 2b, all interrogated artificial defects were detected. In Figure 2a *(i)* the undamaged reference specimen is depicted. Figure 2a *(ii)* shows the inserted teflon defect in the centre of the repaired plate. In the case of Al substrate, damage is easier witnessed when using the colorful palette. In Figure 2a *(iii)* all artificial PTFE flakes between the patch and the mother structure are visible. In Figure 2b *(iv)* the induced damage between the layers of the patch can be readily discerned. A surface defect visible to naked eye is also identified. In Figure 2b *(v)* the triangular Teflon insert is easily unscrambled. Figure 2b *(vi)* depicts the 50% simulated bad adhesion. In this damage configuration thermography was not as successful as for the rest. For this damage concept, oil grease was employed to provoke "bad" adhesion. The thermographs indicate that "bad" adhesion was not traceable. On the contrary it is possible that the majority of the patch/substrate interface remained intact. Poor adhesion is manifested as fade bright areas in the middle of both repaired substrates. In all phase images above the different layers of the patch are visible. Moreover, high consolidation degree of the manufactured patches was observed. It is noteworthy that the thickness of the Teflon defects was equal to 50 μm, lower than in other studies [6]. Although their detection was feasible, reduced thickness additionally hampered the discerning efficiency of the artificial defects.

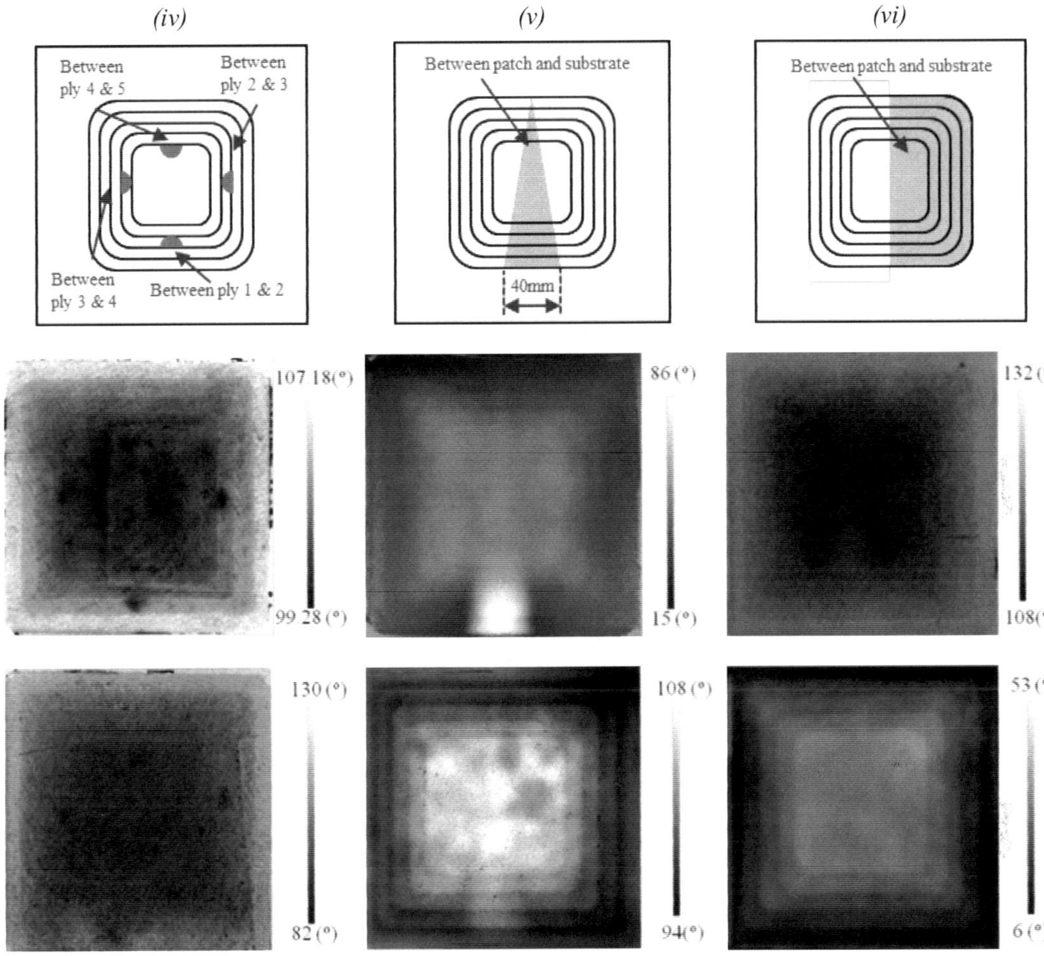

Figure 2b. (A) Coupon configuration, (B) Phase images by lock-in thermography; Al substrate, (C) Phase images by lock-in thermography; Composite substrate, *(iv)* Delamination between patch layers, *(v)* Triangle de-bonding, *(vi)* 50% bad adhesion.

3.2. Service assessment; bonded composite repair of Al helicopter wing

In Figure 3 the testing configuration is depicted. The induced bending moments were in different amplitude in order to equilibrate the distance difference between the centre and the clamped points. The structure was tested in 20Hz cyclic bending fatigue. The infrared camera was synchronized with the frequency of the testing machine providing both amplitude and phase images. The infrared sensor was appropriately positioned vertically-faced above the repaired side of the stabilizer (Figure 3). The distance between the camera and the stabilizer was approximately 0.8m. Having two signal outputs from the controller, the camera was connected with the P1 loading moment. The frame rate of the infrared camera was set to capture four images per loading cycle.

The sequence of amplitude images in Figure 4, were observed with lock-in thermography during representative loading cycles. Amplitude images represent the recorded temperature difference on the structure. The deterioration of the patch is reflected by the color change on the particular images.

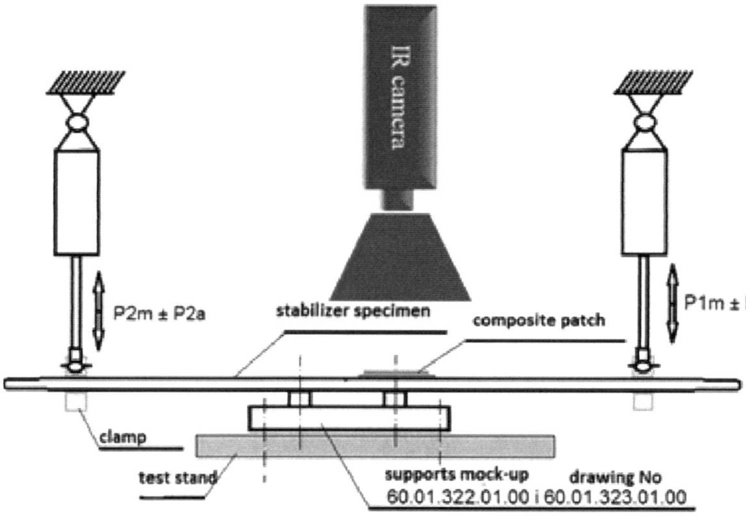

Figure 3. Experimental setup.

The contained data belong to PZL-Swidnik S.A. and any further reprinting and usage cannot be made without PZL-Swidnik S.A. written permission.

4. Conclusions

Within the context of this study, IrT was applied in both maintenance (off-line) and service (online) modes to assess structural integrity of bonded composite repaired structures. Initially, thermography was used to interrogate artificially induced damage simulated by Teflon. Both Al and composite substrates were repaired and subsequently inspected off-line. In a second stage, thermography was employed in real time to monitor initiation and propagation of damage in a repaired helicopter vertical stabiliser.

With regard to off-line investigation, good detection efficiency was achieved for most damage configurations employed in the study. Teflon tapes were identified with higher accuracy than "bad adhesion, induced by oil grease. It is worth mentioning that the applied oil grease is barely visible. This implies that might was inadequate in causing 'bad adhesion', allowing for smooth heat transfer between the patch and the substrate. Detection of defects was less demanding in the case of Al substrate due to the difference in the coefficient of thermal expansion. In the case of composite repaired substrates, higher consolidation between composite-composite as well as the identical coefficient of thermal expansion hindered the detection process. In all cases, Teflon inserts entrapped heat due to their lower thermal conductivity compared to the rest of the structure. Therefore, Teflon was discernible when scanned with a thermal sensor.

In the case of real-time damage assessment, IrT proved a powerful tool for damage initiation and propagation monitoring. The application of load was visualised as different colour contours on the surface of the patch. Mechanical loading generated stresses internally which led to the detachment of the patch from the Al substrate. Changes in colour mirrored the differentiation of the stress gradient along the repaired area. Generated stresses were the result of thermomechanical coupling. Thermographic images reflected the debonding process of the patch from the substrate. Debonding initiated mainly from the loci of the bolted joint where apparently the mechanical stresses were maximum. Stress concentrations moved along the repaired area as the patch detached from the substrate. The artificial crack was initially visible, however, after prolonged testing, the loss of patch /substrate interface covered the thermal trace of the crack. This effect was a secondary proof that IrT was capable of capturing critical damage in real time, which in this study was manifested as patch debonding.

In summary, IrT was successful in detecting damage both in maintenance and service conditions in bonded composite repaired materials and structures.

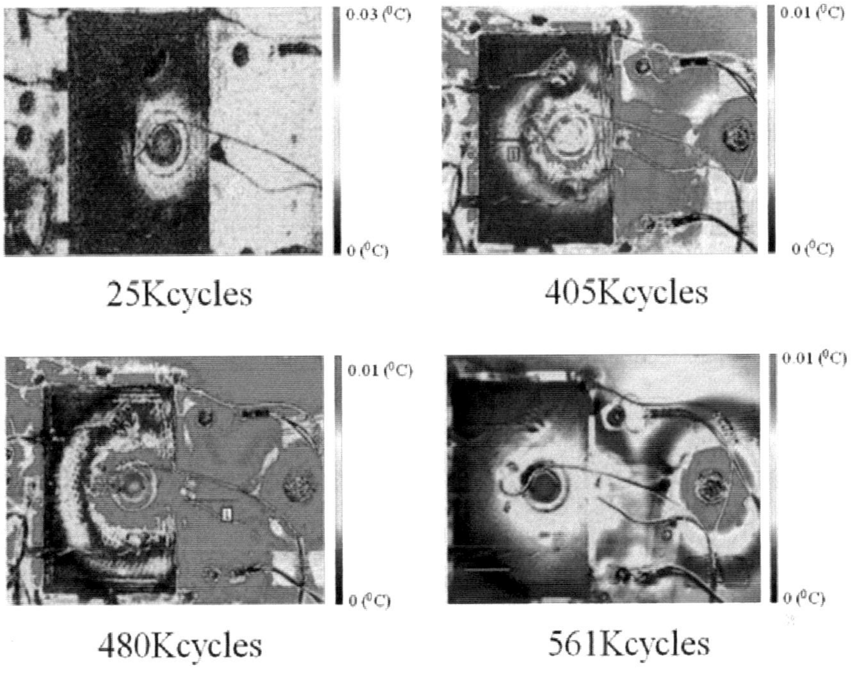

Figure 4. Amplitude images obtained from lock-in thermography at representative loading cycles.

References

[1] Baker AA, 1984, *Composite Structures*, **2** 153-81.
[2] Grammatikos SA, 2013, *Dept. Materials Science Eng.*, University of Ioannina, Greece, 286.
[3] Jasinien E, Raiutis R, literis R, Voleiis A, Vladiauskas A, Mitchard D, Amos M, 2009, *Insight - Non-Destructive Testing and Condition Monitoring*, **51** 477-83.
[4] Grammatikos SA, Kordatos EZ, Matikas TE, David C, Paipetis AS, 2014, *Materials & Design,* **55** 429-41.
[5] Kundu T, Maslov K, Karpur P, Matikas TE, Nicolaou PD, 1996, *Ultrasonics,* **34** 43-9.
[6] Grammatikos SA, Kordatos EZ, Barkoula NM, Matikas TE, Paipetis AS, 2011, *Plastics, Rubber and Composites,* **40** 342-8.
[7] Zweschper T, Dillenz A, Riegert G, Scherling D, Busse G, 2003, *Insight: Non-Destructive Testing and Condition Monitoring*, **45** 178-82.
[8] Pan M, He Y, Tian G, Chen D, Luo F, 2012, *NDT & E International*, **52** 28-36.
[9] Grammatikos SA, Kordatos EZ, Matikas TE, Paipetis AS, 2013, *J. of Materi Eng and Perform*, **23** 169-80.

Analysis of physical mechanical and structural characteristics of microwave cured organic polymer parts

Tatiana Guzeva[1], Yuliya Zavitaeva[1] and Evgeniya Chutskova[2]

[1] Rocket and Spacecraft Composite Structures Department, Faculty Machinery, Bauman Moscow State technical University, 5, 2-nd Baumanskaya Street, 105005, Moscow, Russia.

[2] All-Russian Institute of Aviation Materials, 17, Radio Street, 105005, Moscow, Russia.

nta29@mail.ru, z-yulia@yandex.ru, chutskova@mail.ru

Abstract. This paper presents experimental research which has confirmed the benefits of the new technology of curing polymer matrix for both laboratory and full-scale components, and compares with polymerization in an electrical-heated oven. Strength tests of small size specimens made of organic polymer determined that after microwave heating samples are capable of withstanding 1.3 greater loads than after heating in an electric furnace. Flexural modulus of full-scale specimens of organo-plastic, which were cured in a microwave radiation field showed greater modulus than samples which were polymerized in an electrical-heated oven at 40% and 20% respectively. The microstructure of the samples treated in the electric furnace were found to be porous, inhomogeneous, binding in large portions mixed up with pores spotted, accumulates on the edges and a separate central zones. Polymerization in microwave oven however, gave a microstructure which is more uniform and the binder distributed throughout the volume.

1. Introduction

With regard to practical application it is important to experimentally verify the benefits of microwave curing polymer binders in full-scale specimen compared to the conventional method of curing in electric furnaces.

As is well-known, microwave technologies have a number of advantages compared to traditional methods such as: two and more order reduction of technological processes time, lower energy consumption, improved ecological state and cleanliness at the place of production, possibility of developing new products of better quality.

This article presents experimental confirmation that the quality of the finished product, which were cured in a microwave oven, is higher than that of product which was polymerized in an electrical furnace. Specifically, the microstructure of these samples and also mechanical characteristics are studied.

2. Preparation of experimental samples

In the course of the experiment small-size samples (Figure 1) and organic polymer tubes were wound over the dielectric expander made of cardboard. The original dimensions of full-size component is: length -1200 mm, inner diameter - 85 mm, outer diameter – 94 mm, gauge – 4.5 mm (Figure 2). Specimens were made from aramid fibers Armos 100A with a density of 1450 kg/m^3 and the elastic modulus of 140 GPa, epoxy amine resin with active thinner was used as the polymer matrix.

Curing for full-size tubes was performed on the experimental plant with rectangular microwave cavity equipped with four magnetrons with 2400 W maximum total capacity of and the operating frequency of 2450 MHz. The combination of chamber dimensions, magnetrons layout and a U-shaped thin-walled steel mandrel provided uniform heating across the length and thickness of the tube wall due to retroflection and superposition of electromagnetic waves. The temperature was measured by thermocouples fixed in an isothermal plane at a short distance from the front surface inside the sample.

The heated tube was placed in the central part of U-shaped mandrel with the open side facing to the magnetrons at a distance of several tens of millimeters from the bottom by applying radio-transparent plastic bearing (Figure3).

Figure 1. The received experimental sample.

Figure 2. Winding pipe machine RPN 380.

Besides experimental confirmation of the benefits of microwave curing in comparison with a traditional method for tubes from organic polymer, it is also necessary to compare weight and mechanical characteristics of the tubes made of fibreglass and organic polymer.

Geometrically similar tubes made of fibreglass and organic polymer were produced to compare weights (Figure 4). The weight of organic polymer tube in comparison with a similar fiberglass tube was lower by 24% and amounted to 1.91 kg and 2.52 kg respectively.

Figure 3. Location and spacing of blanks inside U-shaped mandrel microwave heating chamber.

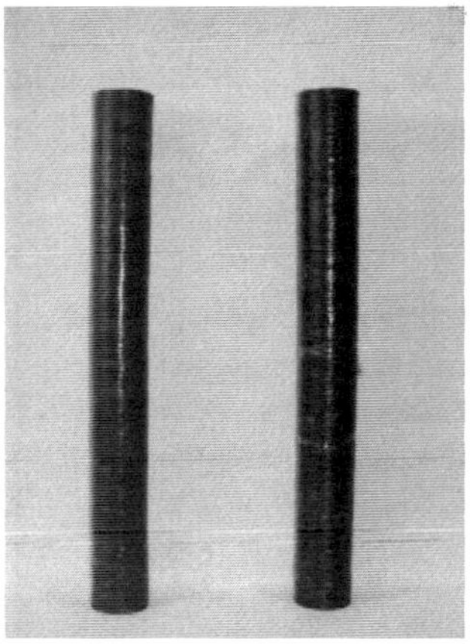

Figure 4. Tubes from organic polymer and fibreglass.

3. Experimental determination of mechanical properties of samples

Strength tests of small size specimens from organic polymer demonstrated (Figure 7) that after microwave heating samples are capable of withstanding 1.3 times greater loads than after heating in the electric furnace. The plant and the test circuit used for the small size specimens are shown in Figures 5 and 6.

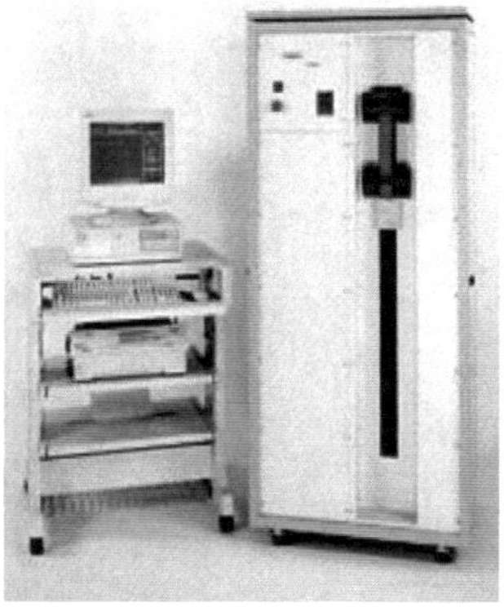

Figure 5. General view of plant STATIGRAF M.

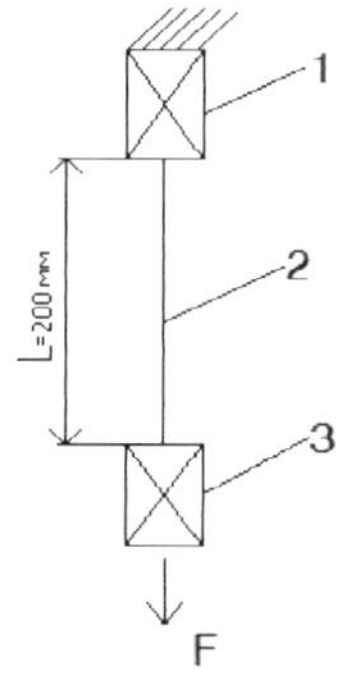

Figure 6. Test circuit: 1 – fixed clamp; 2 – specimen; 3 – movable clamp.

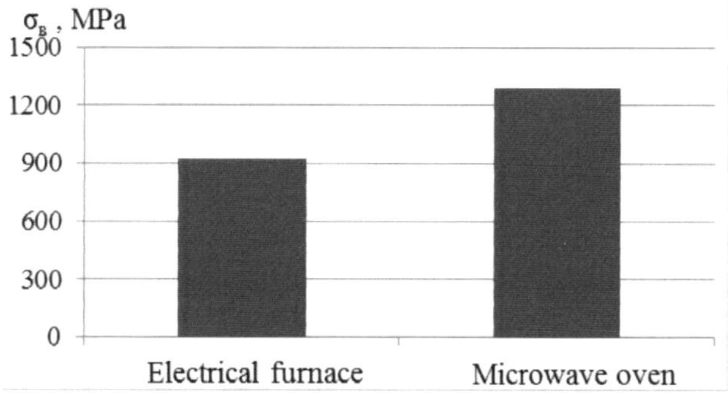

Figure 7. Comparison of durability of organo-plastic part blank.

Figure 8. Bending test.

During the bending tests on the tubes of polymer composite materials (Figures 8 and 9), the amount of change in the diameter along the line of force exertion was determined. The error of measurement was ±1% .

The destruction of specimens from organic polymer continued to a quasi-plastic ball joint formation in a section. For identical fiberglass samples the experiment was carried out before destruction with crack formation.

It was found that the flexural modulus of full-scale specimens of organic polymers, which were cured in microwave radiation field is greater than the same modulus of samples made of fiberglass and organic polymers, which were polymerized in electrical-heated oven at 40% and 20% respectively (Figure 10).

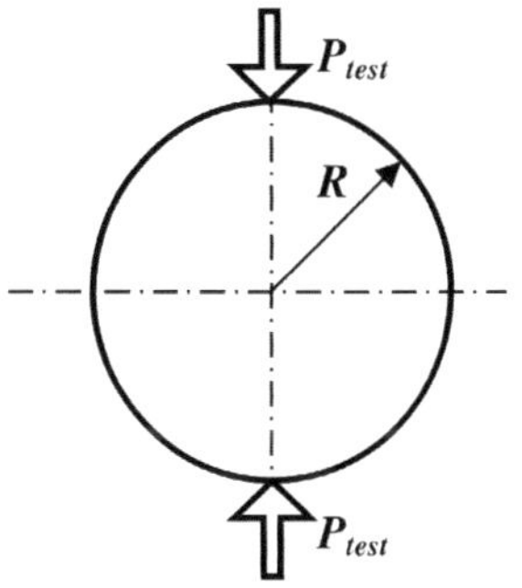

$$E = \frac{0{,}149 P_{исп} R^3}{J \Delta\delta^*}$$

Figure 9. Test circuit.

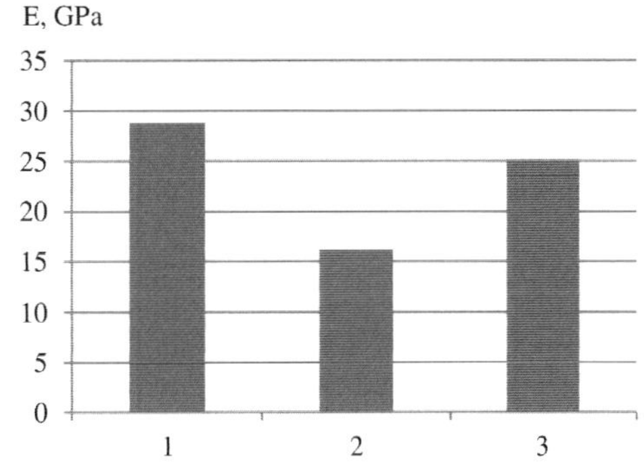

Figure 10. Modulus of elasticity in flexure for experimental samples:
1 – organic polymers, processed by microwave radiation,
2 – organic polymers, processed in an electric furnace,
3 – fibreglass, processed in an electric furnace.

4. The definition of the structural characteristics of organic polymer.

Examination of the microstructure and porosity of samples was performed using inverted-stage microscope of the flat field Neophot-21 working in reflected light.

The microstructure was studied in cross-sections of model samples and natural products from organic polymer.

In Figure 11, 12 results of the analysis of a microstructure of samples from organic polymer are represented. By means of the KSLite 2 program processing of the received results was carried out.

The microstructure of the samples cured in the electric oven is more porous and heterogeneous; the resin is unevenly distributed across the bulk tending to be located at the edges and several large areas in the center. In the case of the microwave cured samples the structure is more uniform and the resin is distributed evenly across the bulk.

5. Conclusions

As a result of the analysis of literary sources it was found that the value of the flexural modulus for the PCM based on fibreglass and epoxy resin, as well as organic fibres and epoxy resin is 25-35 GPa and 15-30 GPa, respectively. After manipulation of the experimental data presented in this paper, the modulus of elasticity is: 28.8 GPa for tube made of organic polymer, which was cured in a microwave oven and 16.2 GPa for the same tube cured in an electric furnace; 25.1 GPa for fibreglass tube, which was cured in an electric furnace. The results show good consistency of the experimental data with the data available in literature.

After testing and calculations it was established that the modulus of elasticity for tube from organic polymer cured by microwave radiation is higher than the modulus of elasticity for the same tube cured by traditional method by 40%. It is also possible to see that the modulus of elasticity of a tube from organic polymer which was polymerized in the field of microwave radiation is higher than the modulus of elasticity of a fiberglass tube which was cured in an electrical furnace by 13%.

All this is confirmation of the benefits of polymerization products of PCM in the chamber of microwave radiation compared to curing of PCM in a conventional furnace. By comparing results of the tubes from organic polymer cured in a microwave oven with fiberglass tubes it was found out that the weight of organic polymer tube was lower by 24%.

The results of structural analysis have shown that in the sample 1 (curing in a microwave oven) the percentage of pore is almost 20 times less than in the sample 2 (curing in an electric furnace), indicating that sample 1 cured more efficiently.

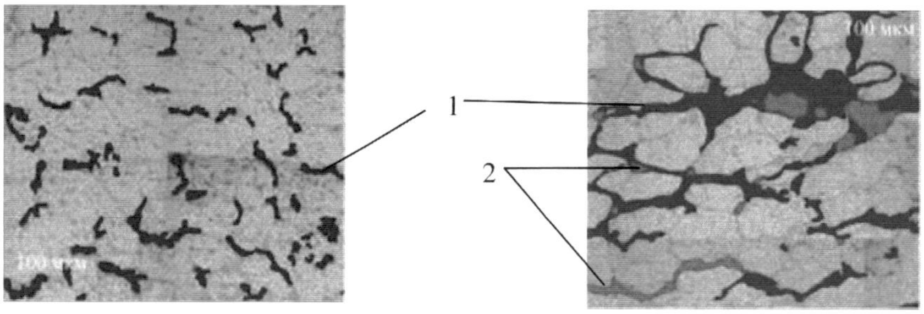

(a) (b)

Figure 11. Microstructure of organic polymers in ring specimens:
a) - treated in a microwave oven, b) - treated in an electric furnace; 1-epoxy binder, 2 – pores.

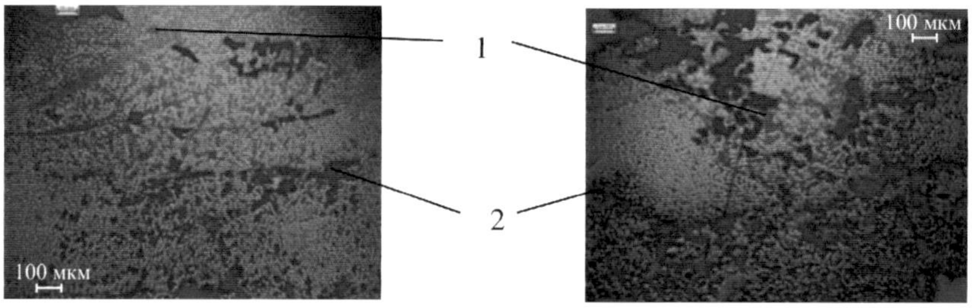

(a) (b)

Figure 12. The microstructure of organic polymers in tubular samples: selection pore spaces and unreinforced binder:
a) - Sample 1 (cured in a microwave oven);
b) - Sample 2 (cured in an electric furnace); 1 - pores 2 - unreinforced binder.

6. References

[1] Day RJ and Samoladas E, 1998, "Micromechanics of model carbon epoxy composites cured using microwave heating." *Science and Engineering of Composite Materials* **7** 23.

[2] Kalganova SG, 2006, "Influence of a SHF electromagnetic field upon kinetics of epoxy-resin curing." *Bulletin of Saratov State Technical University* **1** 90-96.

[3] Mamontov AV, 2008, "Microwave technologies." *Scientific Research Institute for advanced materials and technologies* 308.

[4] Mamontov AV, 2013, "Methods to reduce the level of supplementary emissions from a microwave beam type plants for heat treatment of materials." *EMC Technology* **№3 (46)** 24-28.

[5] Birger IA, Panovko YG, 1988, *Strength. Stiffness. Oscillation. Reference book in three volumes* **5** 831.

[6] Tarnopolsky YM, Kincis TY, 1981, *Methods of static tests of reinforced plastics* (Moscow: Chemistry).

IOP Conf. Series: Materials Science and Engineering **74** (2015) 012008 doi:10.1088/1757-899X/74/1/012008

Production and properties of micro-cellulose reinforced thermoplastic starch

Á. Kmetty, J. Karger-Kocsis, T. Czigány

MTA–BME Research Group for Composite Science and Technology, Muegyetem rkp. 3., H-1111 Budapest, Hungary.
Department of Polymer Engineering, Faculty of Mechanical Engineering, Budapest University of Technology and Economics, Muegyetem rkp. 3., H-1111 Budapest, Hungary.

kmetty@eik.bme.hu

Abstract. Thermoplastic starch (TPS)/micro-fibrillated cellulose (MFC) composites were prepared from maize starch with different amount of distilled water, glycerol and cellulose reinforcement. The components were homogenized by kneader and twin roll technique. The produced TPS and TPS-based polymer composites were qualified by static and dynamic mechanical tests and their morphology was analysed by microscopic techniques. The results showed that the amount of water and the order of the production steps control the properties of both the TPS and its MFC reinforced version. With increasing content of MFC the stiffness and strength of the TPS matrix increased, as expected. Microscopic inspection revealed that the TPS has a homogenous structure and the MFC is well dispersed therein when suitable preparation conditions were selected.

1. Introduction

Nowadays considerable efforts are undertaken to modify biodegradable polymers (*e.g.* thermoplastic starch, polylactide acid) by cellulose micro- and nanoparticles to produce fully biodegradable composite materials with promising mechanical and barrier properties [1-4]. Cellulose is the most abundant natural biopolymer on earth. It is renewable, biodegradable and non-toxic. When extracted from natural fibers, the hierarchical organization of cellulose allows us to obtain micro-fibrillated (MFC). MFC is milled and submitted to bleaching to remove lignin and hemicellulose. The bleached fiber is then hydrolyzed or disintegrated mechanically or both. The resulting MFC is already available commercially by several providers. The diameter of MFC is in the range of 20-60 nm and has a length of several micrometers. Thermoplastic starch (TPS) can be produced from various starches (*e.g.* maize, potato) by the combination of temperature, shear and a plasticizer, which is usually water and/or glycerol. Disadvantages of the TPS are their poor mechanical properties and high sensitivity to humidity which limit their industrial applications.

The goal of this work was to study the effects of plasticizers added in different amount and introduced in different ways. A further aim was to assess the potential reinforcing effect of MFC on TPS (incorporated in various amounts) and clarify the related structure-property relationships. Accordingly, TPS was prepared using water and glycerol by melt kneading in an inner mixer. MFC was introduced in both dry and wet forms. The latter means the introduction of MFC in aqueous suspension. The order of addition of plasticizer and MFC was also tested.

2. Materials and methods

2.1. Materials

As matrix material conventional maize starch (Hungramid F Meritena 100, Brenntag Ldt., Hungary) and as plasticizers different amount of distilled water and glycerol 99.5% (Csepp Bt., Hungary) were selected and used. Two kind of micro-fibrillated cellulose (Arbocell® UF100 (fiber length: 8μm) and B600 (fiber length: 60 μm) from JRS GmbH, Germany) were applied as reinforcements. The moisture contents of the maize starch, UF 100 and B600 MFC, when stored at 24°C, 50% relative humidity, were 9%, 4% and 7%, respectively. These data were received in thermo gravimetric analysis performed at 10°C heating rate in the temperature range: 20-200°C. TPS and TPS/MFC systems were produced under different processing conditions. Table 1. shows the recipes of the materials.

	Thermoplastic starch			MFC reinforcement		
Sample	Starch [wt%]	Glycerol [wt%]	Distilled water [wt%]	Glycerol [wt%]	Distilled water [wt%]	Type of cellulose
Ref	60	20	20	-	-	-
2B	60	20	20	-	-	UFC100
2C	60	20	20	-	-	B600
3B	60	10	10	10	10	UFC100
3C	60	10	10	10	10	B600
5C-5	60	0	0	20	20	B600
5C-10	60	0	0	20	20	B600
5C-15	60	0	0	20	20	B600
6B-5	50	0	0	20	30	UFC100
6B-10	50	0	0	20	30	UFC100
6B-15	50	0	0	20	30	UFC100

Table 1. Recipes and production of the materials.

Coding: Ref: Thermoplastic starch (TPS) reference; 2: TPS produced by introducing dry cellulose; 3: TPS produced by halving the plasticizers for starch destructurization and "wet" introduction of MFC; 5: TPS production when the whole plasticizer amount served to disperse the MFC and introduce it "wet"; 6: TPS produced as sample 5, however, with increased amount of plasticizer

2.2. Preparation of thermoplastic starch based composites

2.2.1. Kneading process. Maize starch was dried at 85°C for 10 hour by KDCL type dry air dryer prior the processing. The components were mixed together with a Brabender type of mixing paddles in different ways. First, the reference TPS was prepared (Ref). The maize starch, glycerol and the water were mixed and fed to the kneading chamber (full amount was 200 g). To produce samples 2B and 2C the dry cellulose was mixed with the plasticized starch. In case of sample 3 series the plasticizer amount was halved: one half was used for TPS, whereas the other to predisperse MFC that was thus introduced in form of MFC suspension. In cases of the series 5 and 6 the whole amount of the plasticizers was used to make an MFC dispersion that was dosed to the dry starch in the kneader. The applied reinforcement content was 5 wt% uniformly, however, samples 5 and 6 were produced also with 10 and 15 wt% MFC contents to shed light on the reinforcement effect. Kneading occurred in two steps: first at 90°C for 2 min at 20 revolutions/min (rpm), and second at 120°C for 8 min mixing time at 20 rpm.

2.2.2. Sheeting process. The homogenized mixture of the components was fed to twin-roll mill (Labtech Scientific Ltd, Thailand) for sheeting. The related conditions were: back roll temperature: 40, 38, 40°C, front roll temperature: 60, 58, 60°C; gap: 0.7-2 mm; friction: 17:15 rpm; tolling time: 8 min. The final thickness of the sheets was 1.6 mm.

2.3. Specimens and their testing
The produced reference TPS and cellulose reinforced TPS sheets were air-conditioned in a Memmert HCP153 humidification chamber at 27°C and 50% relative humidity for 10 h prior to their testing.

2.3.1. Static tensile tests. Static tensile tests were performed on the sheets. From the sheets dumbbell specimens (EN ISO 8256 type 3) were cut off by punching. Tensile tests were carried out on a universal Zwick Z020 tensile machine according to the standard EN ISO 527. The cross-head speed was 5 mm/min and each test was performed at room temperature (24°C). At least five specimens were tested for each material.

2.3.2. Dynamic mechanical analysis (DMA). The DMA tests were performed on a DMA Q800 (TA Instruments, New Castle, USA) machine using three point bending and dual cantilever type loading conditions with the following parameters: frequency: 1 Hz, temperature range: -50 to 120°C, heating rate: 3°C/min. For the tests 60x15x1.6 mm specimens were cut from the sample 5 sheet. Before the DMA tests the surface of the specimens were coated with silicon based grease to avoid/hamper drying during the test [5].

2.3.3. Scanning electron microscope (SEM). Scanning electron microscopic pictures were taken from fracture surface (which was produced by cryogen cracking method) in a JSM-6380LA (Jeol, Tokyo, Japan) microscope. The sample's surface was sputter coated with gold alloy to avoid electrostatic charging.

3. Results and discussion
The primary goal of this work was to find a reproducible production method for TPS based composites which results in optimum properties (maximal strength while maximal modulus). Therefore, first the effects of the plasticizers were investigated. Without distilled water a homogenous, powder-like product was received whereby the kneading torque significantly increased. Omitting the glycerol from the recipe yielded a dry material. It can be thus claimed that the common use of distilled water and glycerol was necessary to produce TPS successfully. The common use of these two plasticizers along with the shearing and heating in the two-step kneading resulted in a dough-like material. Its appearance indicated the good destructurization of the starch. The second mixing step at increased temperature (120°C) caused steaming that was most helpful to the homogenization.

3.1. Static tensile tests
Table 2. shows the mechanical properties of the TPS and TPS based composites. Sample 2 TPS systems exhibited very poor mechanical properties, nevertheless, with increasing amount of the

plasticizers the MFC reinforcing effect was enhanced. Comparing the effect achieved by the two different MFCs, it can be seen that the longer MFC (Arbocell B600, 60 μm) ensured better properties. In addition, the production of the corresponding TPS/MFC composite was easier and the cellulose dispersion became better (cf. Fig.3) than with the other MFC type (UFC 100). For the samples 5 incorporation of MFC in 15 wt% improved the yield stress by 120% and the Young's modulus by 66% compared to the TPS reference.

Sample	Yield stress [MPa]	Young's modulus [GPa]	Strain at F_{max} [%]
Ref	4.2±0.3	0.3±0.02	28.9±2.6
2B	3.4±0.2	0.8±0.03	78.0±2.3
2C	5.2±0.4	0.3±0.01	22.6±2.0
3B	3.9±0.6	0.2±0.06	17.5±5.6
3C	6.3±0.9	0.2±0.03	13.3±3.0
5C-5	5.7±0.5	0.2±0.02	33.6±3.4
5C-10	4.8±0.7	0.2±0.04	27.8±3.6
5C-15	9.4±1.4	0.5±0.06	14.2±2.1
6B-5	2.4±0.3	0.05±0.01	69.3±2.3
6B-10	7.6±0.2	0.3±0.03	31.2±2.2
6B-15	7.9±1.0	0.4±0.05	15.6±3.3

Table 2. Mechanical properties of TPS and TPS/MFC composites

3.2. Dynamic mechanical analysis (DMA)

Figure 1 shows the DMA curves of the TPS based 5C composites which proved to be most promising in this test series (stable production technology, high mechanical properties).

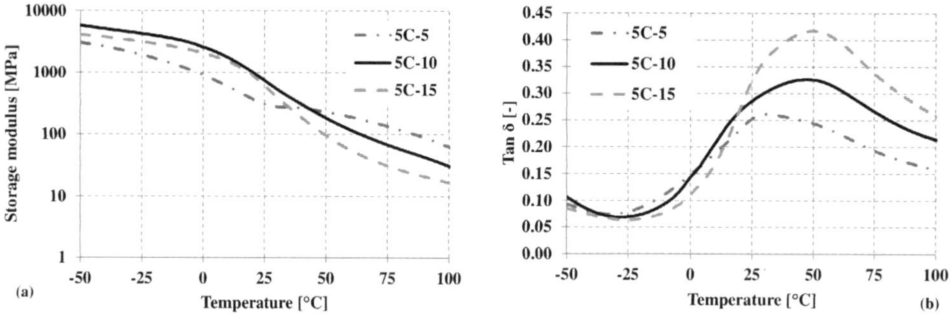

Figure 1. Storage modulus (a) and tan δ curves (b) of the TPS composites (sample 5C), reinforced with different amounts of MFC

As expected from the thermoplastic nature of TPS, with increasing temperature the storage modulus decreased. The glass transition temperature values (read at the maximum of tan δ) of the composites were between 30 and 50°C. This is in accordance with the literature [6].

IOP Conf. Series: Materials Science and Engineering **74** (2015) 012008 doi:10.1088/1757-899X/74/1/012008

3.3. Scanning electron microscopy (SEM)

The SEM pictures from the fracture surfaces of TPS and TPS produced with dry MFC dosage can be seen in Figure 2.

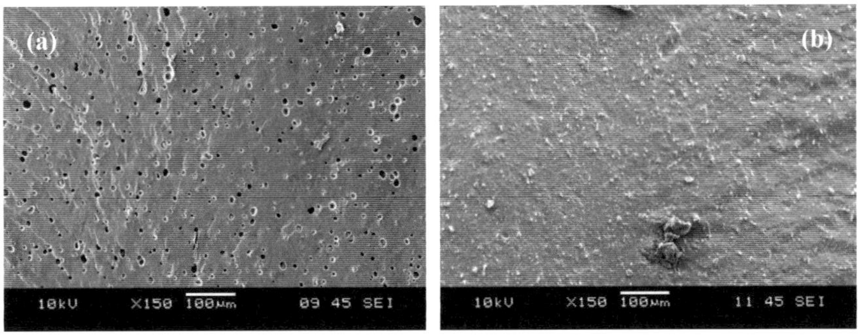

Figure 2. SEM pictures of the reference thermoplastic starch (a) and sample 2B (b).

The TPS was homogeneous but unfortunately contained many bubbles which is certainly decreased the mechanical properties. Their onset can be attributed to the steam development during kneading. When dry cellulose was added to the plasticized starch, no bubble formation could be resolved. Probably the excess water was "picked up" by the dry cellulose. In Figure 3. the fracture surfaces of sample 5 with 5 and 15 wt% MFC contents are collated.

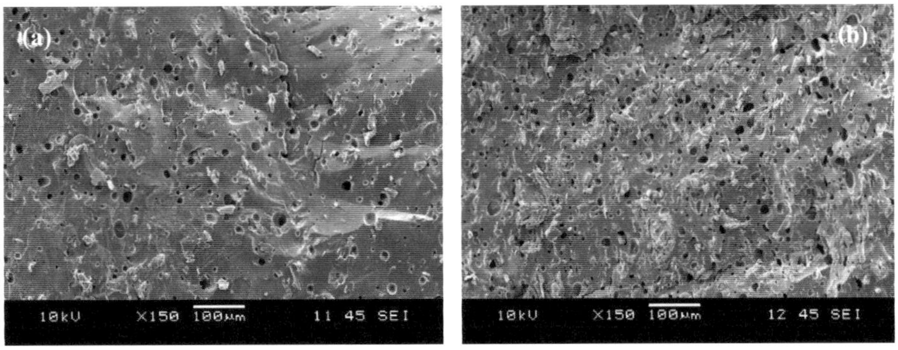

Figure 3. SEM pictures of the 5 (a) and 15 wt% (b) MFC-containing TPS composites (samples 5C-5 and 5C-15, respectively)

The dispersion of the cellulose particles seem to be fine, moreover, MFC is likely fairly adhered to the TPS matrix. Unfortunately, massive void formation due to steam bubbling is still there - its elimination is our major task in the future.

4. Conclusions

In this work TPS and TPS-based micro-fibrillated cellulose (MFC) reinforced composites were produced and tested. It was found that the preparation method and the plasticizers (amount and ratio) strongly affect the resulting properties. Incorporation of MFC increased the mechanical properties (yield stress, Young's modulus, and even ductility). To avoid the agglomeration of MFC it has to be introduced in "wet" form, i.e. predispersed in the plasticizers. The results received will serve as base to produce TPS composites continuously using extrusion melt compounding.

5. Acknowledgements

This research was realized in the frames of TÁMOP 4.2.4. A/1-11-1-2012-0001 "National Excellence Program – Elaborating and operating an inland student and researcher personal support system". The project was subsidized by the European Union and co-financed by the European Social Fund. The research was supported by the Hungarian Scientific Research Fund (OTKA K109409 and K105257) and Hungarian Academy of Sciences (bilateral cooperation between Mexico and Hungary). This work is connected to the scientific program of the "Development of quality-oriented and harmonized R+D+I strategy and functional model at BME" project. This project is supported by the New Széchenyi Plan (Project ID: TÁMOP-4.2.1/B-09/1/KMR-2010-0002).

References

[1] Siquera G, Bras J and Dufresne 2010 *Polymer* **2** 728.
[2] Lavoine N, Desloges I, Dufresne A, Bras J 2012 *Carbohyd.Polym.* **90** 735.
[3] Karger-Kocsis J 2008 *Express Polym. Lett.* **2** 312.
[4] Courgneau C, Rusu D, Henneuse C, Ducruet V, Lacrampe M-F, Krawczak P 2013 Express Polym. Lett., **7** 787.
[5] Rodriguez-Gonzalez F J, Ramsay B A and Favis B D 2004 Carbohyd. Polym.**58** 139.
[6] van Soest J J G and Vliegenthart J F 1997 Trends Biotechn., **15** 208.

Discrete dislocation dynamics simulations in a cylinder

Maosheng Li[1], **Chan Gao**[2], **Jianing Xu**[1]
[1] Applied Physics and Computational Mathematics, Beijing 100088, China
[2] Institute of Nuclear Physics and Chemistry, Mianyang, Sichuan 621900, China

li_maosheng@iapcm.ac.cn

Abstract. Mechanical properties of material are closely related to the motion of dislocations, and predicting the interactions and resulting collective motion of dislocations is a major task in understanding and modelling plastically deforming materials. A discrete dislocation dynamics model is used to describe the orientation substructure within the microstructure. Discrete dislocation dynamics simulations in three dimensions have been used to examine the role of dislocation multiplication and mobility on the plasticity in small samples under uniaxial compression. In this paper we describe the application of the dislocation dynamics simulations in a cylindrical geometry. The boundary conditions for the simulation were estimated from the distribution of the geometrically necessary dislocation density which was obtained from the orientation map. Numerical studies benchmark could validate the accuracy of the algorithms and the importance of handling the singularity correctly. The results of the simulation explain the formation of the experimentally observed substructure.

1. Introduction

Material mechanical properties, such as strength and deformation of crystalline materials, are largely determined by underlying mechanisms involving various crystal defects, such as vacancies, interstitials and impurity atoms, dislocations, grain boundaries, heterogeneous interfaces and micro-cracks, chemically heterogeneous precipitates, twins and other strain-inducing phase transformations. Among all of them, dislocations, which provide a useful link between crystal defects in atomic scales and the macroscopic response of crystals, are the most attractive form of defects for worthwhile study. Most often, dislocations define plastic yield and flow behaviour, either as the dominant plasticity carriers or through their interactions with the other strain-producing defects [1, 2].

Because of the multiplicity and complexity of the mechanisms of dislocation motion and interaction, it is nearly impossible to develop a quantitative analytical approach for understanding and mimicking these problems. The situation is further complicated by the need to trace the spatio-temporal evolution of a large number of interacting dislocations over long periods of time, as is required for the calculation of plastic response in a representative volume element. Such practical intractability of the dislocation-based approaches, together with the developing needs of material engineering at the nano- and micro-length scales, has created the current situation where the equations of crystal plasticity used for continuum modelling are phenomenological and largely disconnected from the physics of the underlying dislocation behaviour.

Fortunately, with the advancement in computational technology with bigger and faster computers, bridging the gap between dislocation physics and continuum crystal plasticity has become possible. In order to mimic the dislocation motions, various discrete dislocation dynamics models have been developed over the last two decades. In the 1990s, development of new computational approaches of dislocation dynamics in three-dimensional space (3D) generated hope for a principal breakthrough in our current understanding of dislocation mechanisms and their connection to crystal plasticity [3-5], however, complications relate to dislocation multiplications, self-interactions and interactions with other defects, and keeping track of complex reactions and mechanisms have provided new challenges for developing efficient computational algorithms. To overcome these challenges, many researchers have been dedicated to devise the dislocation dynamics analysis and its computer simulation modelling over the past decade and they achieved significant progress in dealing with the problems [3, 6]. This progress has been further magnified by the idea to couple dislocation dynamics with continuum mechanics analysis in association with computational algorithms such as finite elements.

Our long term goal is to develop a physics-based multi-scale modelling hierarchy, validated and verified, to address outstanding questions regarding materials microstructure and mechanical properties. Small-scale structures offer opportunities for direct comparison between modelling and experiment at previously inaccessible scales. Discrete dislocation dynamics simulations in 3D have been used to examine the role of dislocation multiplication and mobility on the plasticity in small samples under uniaxial compression [5-8]. The experiments provide data for validation of models, and the models provide a path for new, physically based understanding and prediction of materials behaviour.

The focus of this article is on showing progress in simulation code development, and comparison of the simulation results with experiment to validate the code [9, 10]. In the following, we firstly present the principles of dislocation dynamics analysis and introduce the simulation procedure. Then we show some simulation results and compare to the experiments. Finally, in the conclusion section, evaluations and our long term project objectives are provided.

2. Theoretical Fundamentals of the Method

2.1. Dislocation Equation of Motion

A dislocation is a line defect in an otherwise perfect crystal described by its line sense vector ξ and Burgers vector b. Under loading, dislocations glide and propagate on slip planes causing deformation and change of shape. When the local line direction becomes parallel to the Burgers vector, the dislocation may propagate into other slip planes. This switching of the slip plane, which makes the motion of dislocations three dimensional, is better known as "cross slip", and is an important recovery phenomena to be dealt with in dislocation dynamics.

The dynamics of the dislocation is governed by a Newtonian equation of motion, consisting of an inertia term, damping term, and driving force arising from short-range and long-range interactions. The dislocation dynamics approach attempts to incorporate all of the kinematics and kinetics aspects into a computational traceable framework. It is possible to represent smooth dislocations with any desired degree of realism, provided that the discretization resolution is taken high enough for accuracy. In such a representation, the dynamics of dislocation lines is reduced to the dynamics of discrete degrees of freedom of the dislocation nodes connecting the dislocation segments.

The velocity v of a dislocation segment s is governed by a first order differential equation consisting of an inertia term, a drag term and a driving force vector [11, 12], such that

$$m_s\dot{v} + \frac{1}{M_s(T,P)}v = F_s \quad \text{with} \quad m_s = \frac{1}{v}\left(\frac{dW}{dv}\right) \tag{1}$$

$$F_s = F_{Peirel} + F_D + F_{Peirel} + F_{External} + F_{Obstacle} + F_{Image} + F_{Osmotic} + F_{Thermal} \tag{2}$$

In the above equation the subscript s stands for the segment, m_s is defined as the effective dislocation segment mass density, M_s is the dislocation mobility which could depend both on the temperature T and the pressure P, and W is the total energy per unit length of a moving dislocation (elastic energy plus kinetic energy). As implied by (1), the glide force vector F_s per unit length arises from a variety of sources described in the previous paragraph.

2.2. Discrete Dislocation Equation of Motion

Equation (1) applies to every infinitesimal length along the dislocation line. In order to solve this equation for any arbitrary shape, the dislocation curve may be discretized into a set of dislocation segments. Then the velocity vector field over each segment may be assumed to be linear and, therefore, the problem is reduced to finding the velocity of the nodes connecting these segments. There are many numerical techniques to solve such a problem. For example, a straight dislocation segment s is defined to be bounded by two nodes j and $j+1$. Then within the finite element formulation, the velocity vector field is assumed to be linear over the dislocation segment length. This linear vector field v can be expressed in terms of the velocities of the nodes such that $v=[N^D]^T V^D$, where V^D is the nodal velocity vector and $[N^D]$ is the linear shape function vector [13].

Upon using the Galerkin method, equation (1) for each segment can be reduced to a set of six equations for the two discrete nodes (each node has three degrees of freedom). The result can be written in the following matrix-vector form,

$$[M^D]\dot{V}^D + [C^D]V^D = F^D \tag{3}$$

where

$[M^D] = m_s \int [N^D][N^D]^T dl$	is the dislocation segment 6x6 mass matrix,
$[C^D] = (1/M_s)\int [N^D][N^D]^T dl$	is the dislocation segment 6x6-damping matrix, and
$F^D = \int [N^D]F_s dl$	is the 6x1 nodal force vector.

2.3. Evaluation of Plastic Strains

The motion of each dislocation segment gives rise to plastic distortion, which is related to the macroscopic plastic strain rate tensor $\dot{\varepsilon}^P$, and the plastic spin tensor W^P via the relations

$$\dot{\varepsilon}^P = \sum_{s=1}^{Ns} \frac{l_s v_{gs}}{2V}(n_s \otimes b_s + b_s \otimes n_s) \tag{4}$$

$$W^P = \sum_{s=1}^{Ns} \frac{l_s v_{gs}}{2V}(n_s \otimes b_s - b_s \otimes n_s) \tag{5}$$

where n_s is a unit normal to the slip plane, v_{gs} is the magnitude of the glide velocity of the segment, V is the volume of the representative volume element and $N_s = N_l \times n_s^{(l)}$ is the total number of segments. Another microstructure quantity, the dislocation density tensor α, can also be calculated according to

$$\alpha = \sum \frac{l_s}{V} b_s \otimes \xi_s \tag{6}$$

IOP Conf. Series: Materials Science and Engineering **74** (2015) 012009 doi:10.1088/1757-899X/74/1/012009

This quantity provides a direct measure for the net Burgers vector that gives rise to strain gradient relief.

2.4. Modifications for Finite Domains

The solution for the stress field of a dislocation segment is true for a dislocation in an infinite domain and for homogeneous materials. While, in order to account for finite domain boundary conditions, the method for solving 3D problems involving free surfaces and interfaces developed by Zbib *et al.* is used [1, 14]. In this paper we perform discrete dislocation dynamics simulations of plasticity at small scales, with consideration of the influences from the free surfaces via a boundary-element method. Cross-slip motion of the dislocation was also included, and found to be critical to the modelling of the dislocation behaviour. Examination of details of the dislocation mechanism illuminates many features unique to small samples and points directly to the importance of both the surface forces and cross-slip in understanding small-scale plasticity.

3. Results and discussion

Recently, a study from Uchic *et al.* presented the size effects in compression of $1.0\mu m$ diameter metal samples [9, 10]. Cylindrical pillars with varying radii were machined with a focused ion beam (FIB) from single-crystal bulk samples and compressed by a blunted nano-indenter. This pioneering work spurred similar activities from several groups, with studies on a range of sample sizes, from sub-micron to tens micron. In the study reported here, we use our three dimensions dislocation code "Dis3D" to simulate the mechanical behaviour of Ni single crystals under uniform compression and compare our results to the results of experiments.

For the simulations in this work, the material properties of nickel used were: shear modulus $\mu = 76\ \text{GPa}$, Poisson's ratio $v = 0.31$ and lattice constant $a = 0.35nm$. The dislocation mobility is taken to be $10^{-4}\text{Pa}^{-1}s^{-1}$ in the calculations, and other parameters were from references [5, 9, 10]. To investigate the effects of loading direction, as well as to make a direct comparison with the experimental results of Dimiduk *et al.* [10], we also prepared three 1.0 μm samples oriented in the [2 6 9] direction. For the single-slip case, only the $1/2[101](\bar{1}11)$ slip dislocation system has the maximum Schmid factor (equal to 0.48). For the multi-slip case, eight slip systems are active, each with the same Schmid factor of 0.41, whereas the other four slip systems have zero Schmid factors and are inactive.

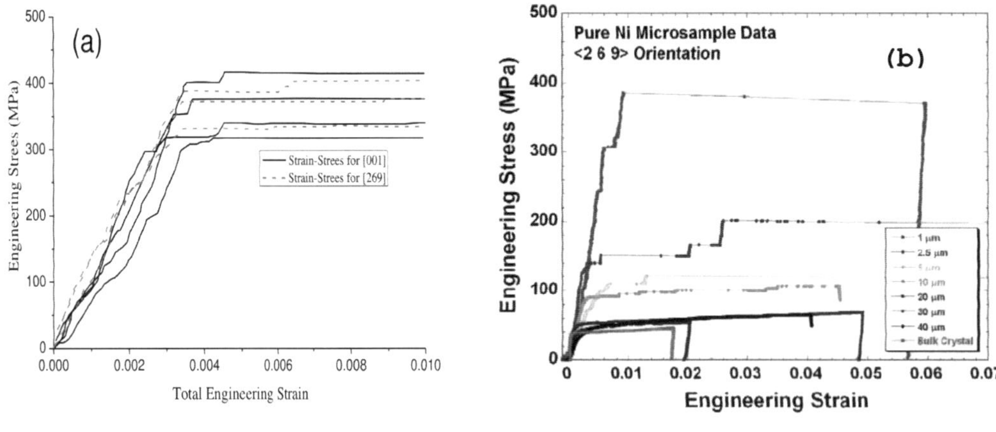

Figure 1. Comparison of stress-strain curves of simulation (a) and experiment [10] (b).

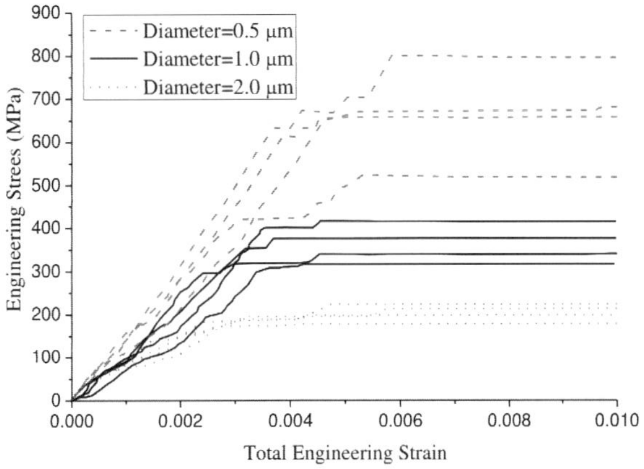

Figure 2. Stress-strain curves obtained from simulation with different sizes.

The stress–strain behaviour for all simulations based on $1.0\mu m$ samples is shown in Figure 1, while the equivalent experimental results for single-crystal nickel are shown. From Figure 1 we can see that the flow stress of the multi-slip simulations (from the [0 0 1] samples) and the single-slip simulations (from the [2 6 9] samples) are both similar to each other and agree well with the experimental results, which employed loading along the [2 6 9] single-slip direction.

In Figure 2 we show a series of stress-strain curves from samples with different diameters under uniaxial compression in the absence of loading gradients. These results illustrate pronounced dependence on size, smaller samples having higher strength.

4. Conclusions

The dislocation dynamics approach developed over the past decade has overcome many hurdles and established a useful framework for predictive simulations. In order to expand the range of engineering and scientific issues that can be addressed, the dislocation dynamics models should become more realistic and computationally efficient. There are still many unanswered questions regarding size-dependent strengthening in small volumes, such as the critical size for transition from bulk behaviour and the role that dislocation structures and mechanisms play in determining that critical size.

In this paper discrete dislocation dynamics simulations in three dimensions have been used to examine the role of dislocation multiplication and mobility on the plasticity in small samples under uniaxial compression. Numerical studies benchmark the accuracy of the algorithms and the importance of handling the singularity correctly. Further investigations are planned for larger samples based on our simulation framework to address these questions. As our research was just a start, we calculate some benchmark to validate the code and to understand the material evolvement procedure. Our goal is to develop a more sophisticated model to predict the mechanical behaviour of microcrystals over a wide range of sizes.

Acknowledgments

This work was performed under the auspices of the National Magnetic Confinement Fusion Science Program in China under contract No. 2010GB111000.

References

[1] Zbib H M and de la Rubia T D 2002 *International Journal of Plasticity* **18** 1133.
[2] Sandfeld S, Monavari M and Zaiser M 2013 *Modelling Simul. Mater. Sci. Eng.* **21** 085006.

[3] Zbib H M, Rhee M and Hirth J P *Advances in Engineering Plasticity and its Applications*, eds. Abe T and Tsuta T. Pergamon, NY 1996:15.
[4] Chakravarthy S S and Curtin W A 2011 *Modelling Simul. Mater. Sci. Eng.* **19** 045009.
[5] Zhou C, Biner S B and LeSar R 2010 *Acta Materialia* **58** 1565–1577.
[6] Weinberger C R Aubry S, Lee S W and Cai W 2009 IOP Conf. Series: *Materials Science and Engineering* **3** 012007.
[7] Aifantis K E, Senger J, Weygand D and Zaiser M 2009 IOP Conf. Series: *Materials Science and Engineering* **3** 012025.
[8] Lee S W, Aubry S, Nix W D and Cai W 2011 *Simul. Mater. Sci. Eng.* **19** 025002.
[9] Uchic M, Dimiduk D M, Florando J N and Nix W D 2004 *Science* **305** 986.
[10] Dimiduk D M, Uchic M D and Parthasarathy T A 2005 *Acta Mater* **53** 4065-4077.
[11] Zbib H M, Rhee M and Hirth J P 1998 *International Journal of Mechanical Science* **40** 113.
[12] Huang H, *et al.* 1999 ASME-JEMT **121** 143-150.
[13] Bathe K J 1982 Finite Element Procedures in Engineering Analysis. New Jersey, Prentice-Hall.
[14] Yasin H, Zbib H M and Khaleel M A 2001 *Materials Science and Engineering* **A309-310** 294-299

Geometry generation challenges for modelling and analysis of micro-structured materials

AJ McMillan

Professor in Aerospace Technology, *Glyndwr University*,
Mold Road, Wrexham, LL11 2AW, UK.

a.mcmillan@glyndwr.ac.uk

Abstract. Engineers evaluating the performance of a component at the design stage will typically convert Computer Aided Design (CAD) geometry into a Finite Element model, and run a Finite Element Analysis (FEA) to determine deformations and stress levels as a result of applied loads or displacements. The analysis results would then be interpreted by comparing them with the required duty of the component. For metallic components, homogeneous and isotropic material properties are generally assumed – "macro-scale" modelling. For components to be manufactured from composite materials, models may represent heterogeneity at the ply level, and orthotropic material properties applied with appropriate directionality. This ply-level modelling is often termed "meso-scale" modelling. Engineering interpretation of failure in materials is often based on empirical understanding of experimental data. This approach is generally robust: safety critical components would always be subject to validation by means of a suitable programme of testing. The aspect that is missing is the opportunity to improve understanding of the material performance by investigating the material performance at the "micro-scale". This paper describes computational algorithms for generating random geometries exhibiting similar characteristics to those seen at the "micro-scale" in real materials, and the use of these to predict the influence of the "micro-scale" structure on the "macro-scale" material performance.

1. Introduction

Almost all materials exhibit some heterogeneity at some length scale: for example, the crystal grain structure in metals, or the individual filaments of fibre in a fibre reinforced composite. Such details are known to play a role in the material performance: for example the strength and ductility of a metal is highly influenced by the crystal grain size and configuration.

The purpose of this paper is to describe the modelling challenges presented by the representation of the micro-structural heterogeneity, and to present results that display remarkable or unexpected details. The mechanical analysis technique adopted here is finite element method [1]. For example, the lower right hand image in Figure 1 illustrates the formation of stress banding in a material containing randomly arranged holes [2].

2. Brief literature review on geometry generation

Geometry creation seems to be a melting pot of disciplines, with relevant literature to be found not only in journals of engineering and applied mathematics, but also in text books and web guides for computer generated art and computer games [3, 4, 5].

IOP Conf. Series: Materials Science and Engineering **74** (2015) 012010 doi:10.1088/1757-899X/74/1/012010

Within the engineering literature, much of the geometry creation work is largely focussed on the modelling of fibre and resin geometry within composites. [6, 7]. Methods of modelling uncertainties in geometry, for example to represent the influence of design tolerance allowables on component strength, are at an early stage of development [8, 9, 10], but the modelling of random features or flaws does not seem yet to have been systematically studied.

3. Porosity in materials

Material porosity is a particular issue for many manufacturing processes, because it is known, for example [11], that porosity reduces the strength of the manufactured component. If a material has holes in it, does it make it weaker, and if so, by how much? Although this seems to be a very simple question, it is rather difficult to specify in a way that can be answered in a quantifiable manner concerning the strength of the component.

On a production line, a manufactured part exhibiting more than a prescribed percentage of porosity will be scrapped. In general, the actual porosity is not measured directly – a direct and accurate measurement system would not be cost effective. In a similar way, the strength reduction resulting from a given level of porosity is not measured in the as manufactured component, but arrived at from previous experience. It would seem that better understanding of the influence of manufacturing process variables and the particular component geometry could lead to a more predictive understanding of what would constitute an unacceptable level of porosity.

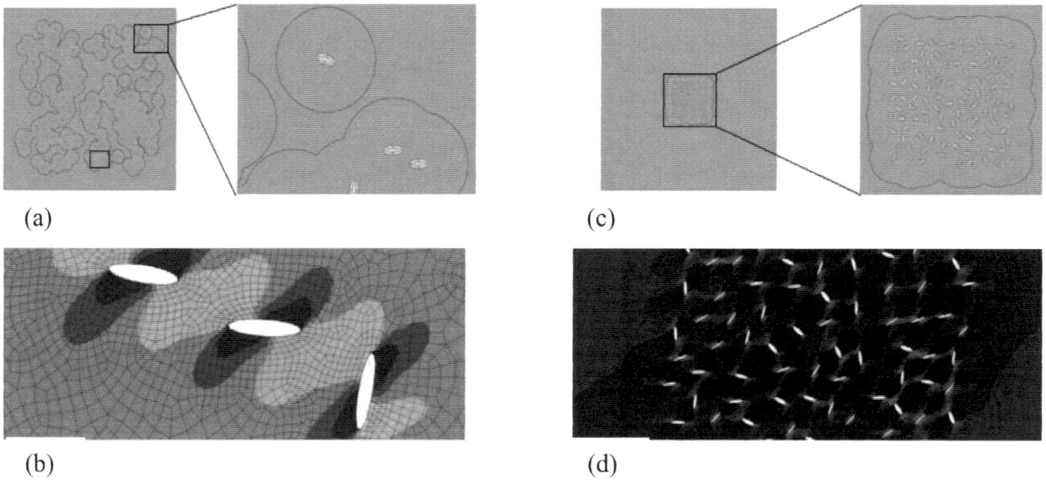

Figure 1. A material filled with elliptical holes; loose packing (a, b) and closely spaced (c, d). The upper Figures (a, c) show the two model domains. The mesh quality is controlled automatically using multiple partitions, indicated by the grey lines; and the yellow regions indicate regions of identical mesh configuration. Figure (b) shows the resulting stress pattern in the vicinity of the three particularly closely spaced holes indicated by the blue rectangle in (a), with the mid blue representing the nominal stress. In (c, d) the packing arrangement is much tighter, but the stress levels (d) are not particularly increased because all the holes are a similar distance from each other. Here the boundary between blue and black regions represents the nominal stress. The stress pattern forms bands and the holes tend to shield each other from extreme stresses.

The model presented in Figure 1 represents a very simplistic initial investigation [2]. A square of material was modelled with Abaqus finite element software, and filled with elliptical holes placed at random but prevented from touching or overlapping. This process was automated using a Python script to create geometry, ensure the proximity conditions, and then generate the finite element mesh and initiate the analysis. The construction curves shown in the pink diagrams help to define and control the topology of the finite element mesh. The square was then subjected to a shearing load to simulate a realistic stress state. In this initial study two forms of packing were defined.

A conclusion from this particular study was that maximum stress increases very slowly with the number of holes, so long as the holes are distributed in a homogeneously random manner. The strength reduction calculation was based on maximum stress values, but control of the mesh around the region of the hole was essential, to ensure against mesh influence [2]. Since the geometry generation algorithm has a major influence on the result, a study of random geometrical forms and a means for categorising them is essential for further development of this field.

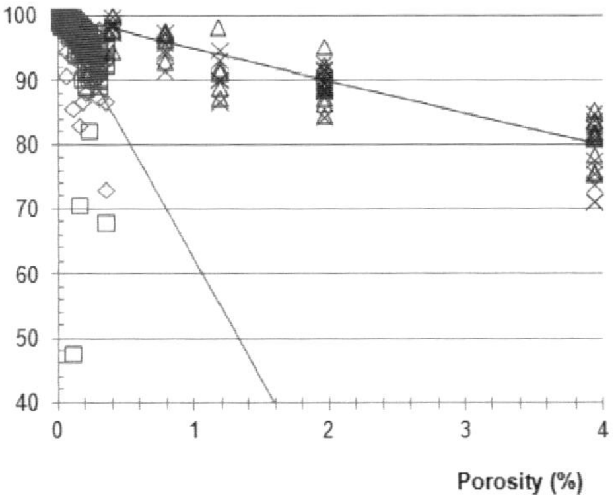

Figure 2. This graph indicates the percentage strength reduction as a result of percentage porosity. There are two distinct sets of data: the essential difference is that where the proximity between pores is more carefully controlled (blue triangles and crosses) a much higher porosity level can be tolerated than in the case where the porosity was more random (red rectangles and diamonds). Note that the red best fit line has been extended beyond the region of applicability, in order for it to be visible.

4. Modelling materials with quantified heterogeneity

Following the initial study, it became clear that a material modelling capability was needed that provided a measurable level of "heterogeneity". Finding an appropriate measure for heterogeneity is a significant part of the problem; nevertheless this has the potential to reveal physical understanding of phenomena that until now have been modelled on the basis of empirical observation.

To put this into context, until now, in component stress analysis, engineers have been obliged to assume material homogeneity, and to adopt pragmatic models for such phenomena as plastic flow, fracture mechanics, and creep. Developing a modelling methodology that can represent a more realistic level of heterogeneity is the starting point to discovering a physical basis for many of the material behaviour rules that are currently determined empirically, and characterised by test.

Ideally, results obtained this way should change the engineering perspective on materials characterisation, but this represents a major undertaking. The obvious challenge is to create the methodology: the less obvious, but potentially far greater, challenge is to persuade the engineering fraternity that such an approach is credible and has the potential to reduce testing costs and lead to better understanding of material performance.

4.1. Empirical vs. real-physics model geometry creation

The challenge is to create realistic geometry for the representation of heterogeneous materials. There are three fundamentally different approaches to generate such geometry. The direct approach is to take a real physical artefact, and generate geometry by measurement. The development of CT imaging capability and the image processing tools means that this is now a genuinely feasible approach [12].

Alternatively, geometry can be created empirically, using CAD drawing tools and lots of imagination. There are two major difficulties here: firstly finding a methodology that can produce realistic looking geometry automatically, and secondly having a algorithmic method for characterizing and validating the "empirical" geometry in comparison with "real" geometry.

A third method is to make use of software embodying relevant physical models, and use the software to develop the model. For example, in Abaqus, a set of hollow spheres could be created, either as elastic solids, or more simply as rigid surfaces, with appropriate contact rules. These could be modelled falling into a box. The physics of contact and mass under gravitational load would ensure that the spheres pack in a realistic way, where the sides of the box arranged to avoid the possibility of perfect packing. Geometry tools could then be used to form a solid, with the insides of the sphere representing the porosity.

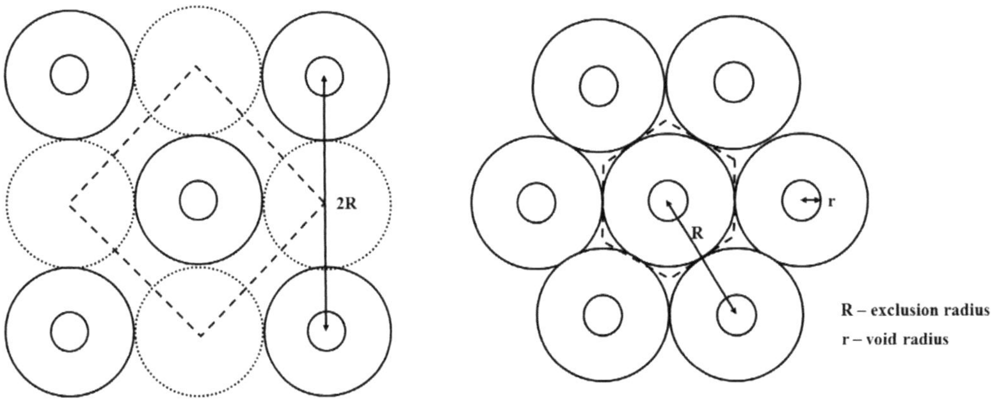

Figure 3. 2D Poisson disc minimum and maximum packing configurations, respectively.

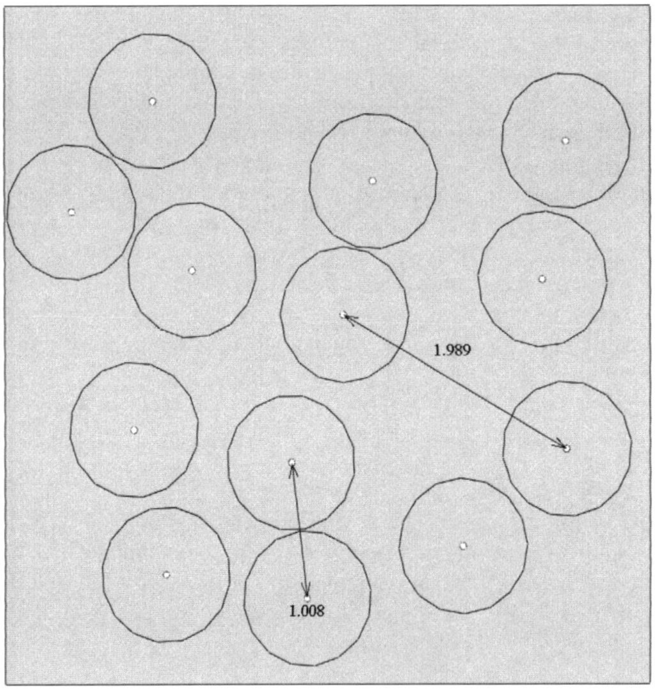

Figure 4. A random porosity model generated in Abaqus using Python. Notice that, even for a small number of seed points, it is not possible to predetermine the number necessary to obtain a "fully dense" packing. The measurements given in this figure demonstrate that the sparse regions are indeed too small to accommodate an additional seed, and the fully dense packing can be verified by inspection.

IOP Conf. Series: Materials Science and Engineering **74** (2015) 012010 doi:10.1088/1757-899X/74/1/012010

4.2. Modelling porosity
Currently, the approach adopted here is to generate 2D geometries using empirical methods. A random number generator is used to define "seed" points throughout the domain, onto which geometrical features are then built. The distribution of the seed points [13, 14] is controlled using the Poisson disc method [5].

The Poisson disc exclusion principal is illustrated in Figure 3. The two configurations represent minimum and maximum packing densities, subject to a rule that no seed point can be closer to another than by an amount R, and that every available space is filled – "fully dense packing". Figure 4 shows a typical result of a Python implementation of a Poisson disc algorithm in Abaqus.

For larger numbers of seed points this scheme is no longer feasible. This is a convenient point at which to introduce the notion of "partial packing". This may be quantified in terms of the ratio of actual seeds placed to the number of seeds required for either the theoretical maximum or minimum fully dense packing. To illustrate this, Figure 5 shows four different packing configurations, each representing the same number of seeds but for different values of R. In this case, the seed points define the centres of small pores in a material, and a nominal biaxial stress field has been applied creating a shearing deformation. The colours indicate relative stress levels [14].

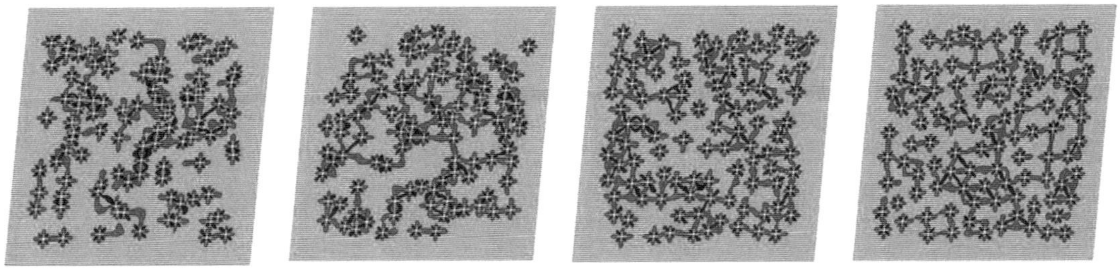

Figure 5. Stress distribution in four models with identical porosity, but with increasingly homogeneous distribution from left to right. Notice how the stress pattern distribution becomes more homogeneous as the seed point distribution tends towards fully dense.

5. Modelling crack propagation in porosity clusters
While single crack propagation using Abaqus XFEM is now a well-established capability [15, 16], multiple crack growth in a porosity cluster is yet to be understood.

Here a problem domain is defined as illustrated in Figure 6. A Python script is written to generate the geometry and partitioning, to enable automatic meshing. Unit traction loads are applied at the boundary to create a biaxial stress state with principal stresses aligned to the diagonals.

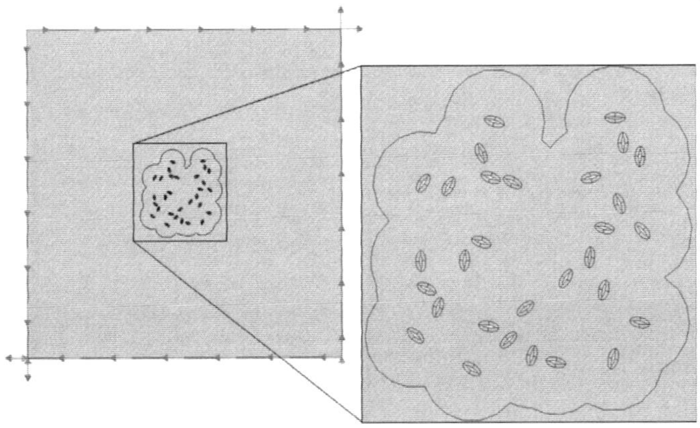

Figure 6. A 2D square of resin material, with a central cluster of randomly positioned elliptical voids.

In Abaqus analysis the loads are ramped up until conditions to initiate cracks are met. Stress levels in some regions are similarly high, and while XFEM is a powerful tool, some care must be taken to ensure all relevant cracks are initiated. XFEM analysis is based on a mesh free enrichment of the domain in the region of the developing crack, and is therefore mesh independent. At each increment, the domain around the propagating crack is re-meshed, nevertheless varying the mesh is sufficient to highlight a problem with sensitivity of crack initiation condition threshold (Figure 7), and raises a question regarding mesh dependency. While such small modifications might influence actual crack path, taken on balance, do the cracks modelled resemble reality?

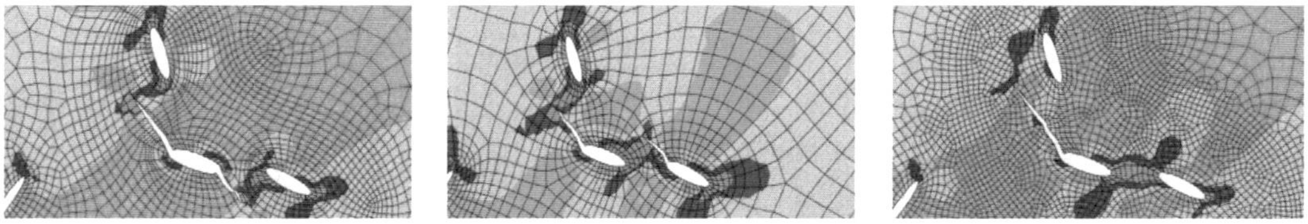

Figure 7. Mesh sensitivity in crack initiation and propagation prediction.

6. Modelling heterogeneous polycrystalline materials

In modelling polycrystalline material structure the same seed point approach can be used. For example, this may mimic the seeding of crystal formation in the solidification of a metal. Again a Poisson disc algorithm can be used to provide a quantified level of "heterogeneity". The challenge here is to find a method for generating realistic crystal boundaries.

While the solidification seed point of the crystal can be modelled by the Poisson disc random seed point together with a random crystal orientation, the individual crystal boundaries could be defined by Voronoi tessellation based on the set of seed points. Thus each Voronoi cell represents a single crystal with randomly prescribed orientation.

This is equivalent to assuming that each crystal in a domain of material starts to solidify at the same time, at the same rate, and isotropically. In reality, different regions of material in a casting mould will cool at different rates, and crystal growth would be anisotropic, influenced by temperature gradient, crystal orientation, and crystal species.

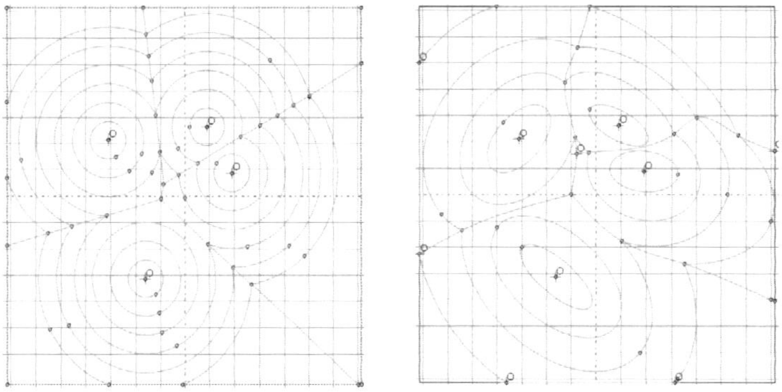

Figure 8. Voronoi tessellation (left) and a modified tessellation for anisotropic boundary growth and time varying growth rate (right). Notice that in this case, the boundaries between crystals are not straight lines.

The most efficient method of creating a Voronoi tessellation is Fortune's algorithm [3, 17], but Figure 8 illustrates a growth based method.

7. Direct modelling of elastic wave propagation

As an alternative to direct geometrical modelling and interpretation of measured geometry, it is possible that the physics of elastic wave propagation through a porous or heterogeneous material could provide results which could be interpreted to provide a measure for porosity or heterogeneity.

Explicit Finite Element is capable of capturing dynamic events, where the elastic wave frequency that can be modelled is limited by the element size, which must be an order of magnitude smaller than the wavelength of vibration.

Thus, by choosing appropriate element sizes and making appropriate modelling choices, it is possible to verify the classical mathematical predictions of wave propagation [18], or to run an analysis on a structure with appropriate initial conditions to ensure vibration is initiated, but with sufficient duration that the standing waves form (Figure 9).

Figure 9. A simulation of high frequency elastic wave propagation, with wave deformation visualised on the external surfaces of a solid block with a central cylindrical hole [19].

In principle, a modelling process such as this could be used to simulate ultrasound based Non-Destructive Test results. Used in tandem with shape optimisation methods, this could provide a technique for enhanced interpretation of results for improved prediction and quantification of flaw size, distribution and location.

8. Discussion – going beyond the current finite element formulation...

In addition to the geometry and modelling challenges presented above, there are two further areas which are worthy of particular consideration.

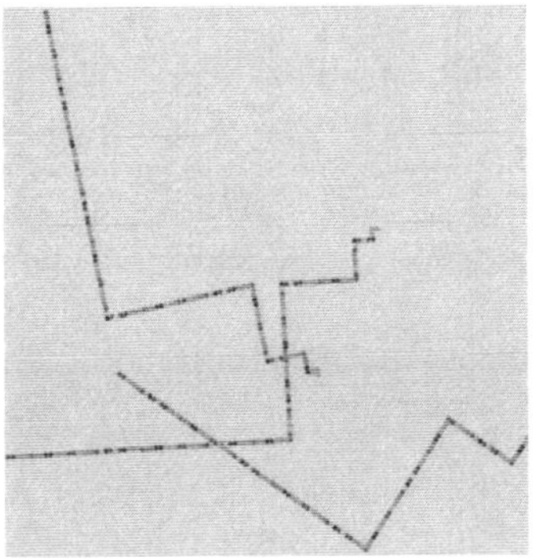

Figure 10. This figure was constructed in Abaqus, using Python, and identifies a fractal path comprised of incremental particular coordinate points within an analysis model. Following analysis, these particular points could be interrogated, taking significantly less time than a review of a full 3D search of the solution domain.

8.1. *Post-processing –searching and manipulating analysis output data*
Routine linear elastic time-independent analysis of homogeneous material domains provides analysis results in which the stress maxima are conveniently located on the surface of the component. As analyses become more sophisticated, it becomes rather more difficult to establish which results are critical and should be assessed. The idea illustrated in Figure 10 is to use 1D fractal based search paths to explore the 3D results domain [14, 20, 21].

8.2. *Solvers which reflect geometrical uncertainty*
In addition to the assumption of material homogeneity, engineering component stress analysis is invariably based on as-designed, or minimum tolerance geometry, but where an as manufactured high value component departs from design allowables it can be time consuming for it to be re-analysed.

Radically different computational solvers might present an alternative approach to standard finite element method, with approaches such as sub-structuring, Mesh-free methods and XFEM already main-stream, but other formulations could provide the key to fast geometry update and re-analysis, as well as critical stress location identification [22].

9. Conclusions
This paper presents a range of challenges presented by materials with heterogeneous properties at the micro-scale. There are a number of open questions for which the development of better computational geometry creation capability would provide useful functionality. Particular lessons already learned are that the length-scales of heterogeneous features, and their orientation and distribution, will have an impact on the macroscopic material properties. It is likely that improved representation of microscopic features could lead to a phenomenological understanding of material behaviours, such as fracture mechanics and creep, which are currently modelled on a macroscopic empirical basis.

Acknowledgements
The initial part of this work was supported under a Royal Society Industry Fellowship.

References
[1] Abaqus Theory Manual, Dassault Systèmes Simulia Corp., Providence, RI, USA.

[2] McMillan AJ, "Material strength knock-down resulting from multiple randomly positioned voids. Journal Reinforced Plastics and Composites", 2012 31(1) 13-28.

[3] M de Berg, M van Krefeld, M Overmars, O Schwarzkopf, "Computational geometry: algorithms and applications", Springer, Berlin, 2000.

[4] Bohnacker H, Gross B, J LaubJ , Ed. Lazzeroni C, "Generative design", Princeton Architectural Press, New York.

[5] Lagae A, Dutré P, "A comparison of methods for generating Poisson disk distributions", Computer Graphics forum, 2008 27(1) 114-129.

[6] Young C, McMillan AJ, Ravey E, Verdicchio J, Quinn J, McIlhagger A, Buchanan S "The hybrid approach of a 3D textile composite finite element modelling technique at meso-scale level" 18th ICCM, South Korea, 2011.

[7] Potter E, Pinho ST, Robinson P, Iannucci L, McMillan AJ "Mesh generation and geometrical modelling of 3D woven composites with variable tow cross-sections" Computational Materials Science 2012 51 103–111.

[8] Matthies HG, "Stochastic finite elements: Computational approaches to stochastic partical differential equations", Journal of Applied Mathematics and Mechanics, 88(11), 849-873, 2008.

[9] Moens D, Vandepitte D, "A survey of non-probabilistic uncertainty treatment in finite element analysis", Computer Methods in Applied Mechanics and Engineering, 194(12-16), 1527-1566, 2005.

[10] Ainsworth M, Oden JT, "A posteriori error estimation in finite element analysis", Wiley-

Blackwell, New York, 2000.

[11] de Almeida SFM, Neto ZSN "Effect of void content on the stregth of composite laminates", Composite structures 28, 139-148, 1994.

[12] Maire E, Withers PJ, "Quantitative X-ray Tomography" International Materials Reviews 2014 59 1-43.

[13] McMillan AJ, Archer E, McIlhagger A, Lelong G, "Strength knock-down assessment of porosity in composites: modelling, characterising and specimen manufacture", Journal of Physics: Conference Series, 2012 382 012027.

[14] McMillan AJ, "Defect identification and characterisation algorithms for assessing effect on component strength", 15th ECCM, Italy, 2012.

[15] Giner E, Sukumar N, Tarancon JE, Fuenmayor FJ, "An Abaqus implementation of the extended finite element method. Engineering Fracture Mechanics", 2009 76(3) 347-68.

[16] Pais M, Web tutorials on XFEM, http://www.matthewpais.com/ [accessed 4 October 2014]

[17] Fortune SJ, "A sweepline algorithm for Voronoi diagrams", Algorithmica, 1987 2 153-74.

[18] Achenbach, JD, "Wave Propagation in Elastic Solids," North-Holland, Amsterdam, 1990.

[19] McMillan AJ, "Computational modeling of ultrasound propagation using Abaqus explicit" SIMULIA UK RUM, Crewe, UK, 2013.

[20] McMillan AJ (Inventor), Rolls-Royce Plc (Assignee), "Method for locating artefacts in a material", United States Patent US 8,774,497 B2, 8 July 2014

[21] McMillan AJ (Inventor), Rolls-Royce Plc (Assignee), "Method for locating artefacts in a material", UK Patent GB2493735B, 23 July 2014

[22] McMillan AJ, Watson G, "The use of wavelets to facilitate finite element post-processing", Materials Science Forum, 2003 440 429-38.

IOP Conf. Series: Materials Science and Engineering **74** (2015) 012011 doi:10.1088/1757-899X/74/1/012011

Discrete action micro-actuator optimization

A Nikolaeva[1], **AJ McMillan**[2], **S Gavriushin**[1]

[1] Bauman Moscow State Technical University, 5/1, 2nd Baumanskaya, Moscow, 105005, Russia.
[2] Glyndwr University, Mold Road, Wrexham, LL11 2AW, UK.

nikolaeva@bmstu.ru

Abstract. Micro-electro-mechanical systems (MEMS) devices are widely used in industry. The micro-actuator is an important component of such devices, transforming external influence into mechanical displacement. The development of a design technique to derive the optimal design parameters for a micro-actuator is a problem of current interest. The present paper describes a technique for determining geometric parameters for a simplified micro-actuator structure (a hemispherical shell), such that under a prescribed critical pressure it will undergo a specified discrete deflection. Such a deflection is commonly referred to as "snap-through". A mathematical model and a Finite Element procedure for the mechanical analysis of a flexible thin-walled shell under large deformation are proposed. Initially the snap-through is modelled as a quasi-static effect, but subsequently, the influence of the inertia is also considered. The optimization procedure was performed using the PSE/MACROS optimization program. The results for an example model numerical optimization are shown.

1. Introduction

The development of innovative base elements for micro-electro-mechanical systems (MEMS) devices has significant potential for the technological modernization of industry. The development of new MEMS device designs is predicated on the development of methodologies for analysis and synthesis of constitutive elements, each with given specific performance characteristics. Each MEMS device comprises a sensor sub-system, a sub-system for analysis and decision making, and an actuator sub-system [1]. This paper focusses on the mechanical aspects of the actuator sub-system.

For a MEMS device to operate efficiently and effectively, the mechanical performance of the actuator sub-system has to be finely tuned. For efficiency, the actuator must be low inertia, and be capable of large motion for low input energy. It must also be capable of moving from one state to another, with minimal lag, and avoid lingering in an intermediate state. For this reason, elastic snap-through structures are ideal: such structures have two (or more) stable states, and only a small applied force is needed to trigger a change of state. A contact lens for vision correction is an example of an everyday object with this kind of performance. Mechanical analysis of such structures is not without its complications, because a small variation of applied load does not lead to a simple small variation of deformation: the structural performance is not linear. This type of non-linear behaviour is known as "non-linear deformation".

There is a well-established theoretical foundation in the field of non-linear deformation of thin-walled structures. The non-linear behaviour of thin-walled shells and snap-through has been documented by many authors [2-8]. More recent research papers [9-10] describe computational algorithms for non-linear deformation. In combination, this development of technology has provided new avenues for research, including the optimization of actuator sub-systems for discrete action MEMS devices. The utilisation of a flexible thin-walled shell element, made of novel materials [11, 12] (*e.g.* silicon), as an actuator is thought to be innovative.

In the present paper, a technique is presented, for determining geometry parameters for a discrete action micro-actuator, operating under snap-through. To keep things simple, the geometry is a hemispherical shell of constant thickness. Note that the radius of the support circle remains constant as it is specified by the device. The parameters in this case are hemispherical radius and shell thickness, subject to particular pressure application conditions and a requirement to achieve a predetermined deflection. In the first instance, the study assumes the snap-through to be a quasi-static effect, but subsequently, the influence of the inertia is also considered. A mathematical model, based on the Reissner thin-walled elastic shell theory [7], for the analysis of flexible shell structures under large deformation was implemented in computer code written by the author, and validated in ANSYS finite element software. A parameter subspace changing method, developed by Gavriushin [13], based prior work on computational algorithms for shells by Valishvili [14], was used to solve the non-linear problem [15]. The optimization was performed using the PSE/MACROS optimization program.

2. The method adopted

The hemi-spherical thin-walled micro-actuator is designed to perform mechanical switching. Let us suppose that its deflection is required to take a given value d^* when the critical pressure P^* is applied. The elastic characteristic of the micro-actuator which includes a snap-through mechanism and the deformed shape of the actuator are shown schematically in Figure 1.

The shell straightens after applying pressure. This part of the process is illustrated by the DA part of the curve. The shell instantly changes the direction of the deflection at the point A (the upper critical point) – the snap-through at point A. Then the shell continues its deformation from the point B of the stable part of elastic characteristic. The deflection of the shell decreases if the pressure decreases (BC) until the point C (lower critical point) and the CD snap-through.

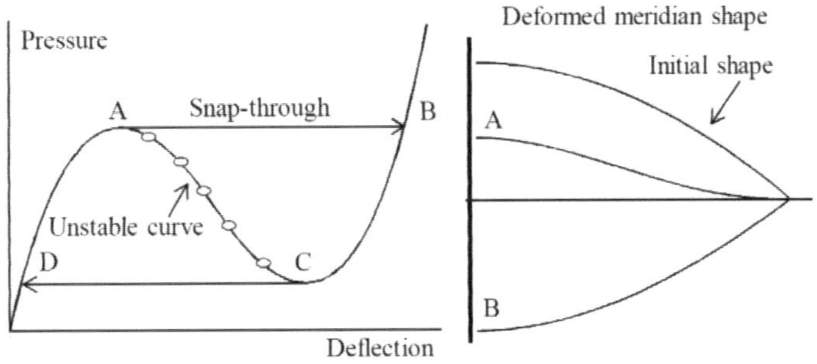

Figure 1. Schematic of Micro-actuator discrete elastic response and deformed shape.

The initial data for the research is given in Table 1. The required outcome of the optimisation process is the selection of values R, hemi-sphere radius of curvature, and h, thickness, subject to the values of critical pressure P^* and the corresponding deflection d^* being equal to customer requirements. The acceptable ranges of variation of these parameters are given in the Table 1. The functional relationships between the criteria and parameters are not set explicitly, but are calculated using Finite Element [16, 17, 18], in ANSYS as described below.

The problem is solved using the axisymmetric element formulation. The shell is modelled using PLANE182 elements. These elements have 4 nodes with 2 degrees of freedom at each node. The material of the shell is linearly elastic. The shell is simply-supported around the edge and is uniformly loaded by pressure. The analysis is performed using the parameter continuation algorithm arc-length method for a problem with large displacements [19, 20]. In this method the length of the arc along the elastic characteristic curve is used as a parameter in the parameter continuation method, rather than displacement (or pressure) in displacement- (or force-controlled modes). This makes it possible to overcome calculation difficulties. The PSE/MACROS program is used for the solution of the optimization problem [21].

In the flowchart detailing the algorithm (Figure 2), the Optimizer block contains information about the parameters, constraints and objective functions. This block governs the solution process.

There are two objective functions, one in terms of applied pressure and the other in terms of snap-through displacement. These were specified as

$$F_1 = |P - P^*| \qquad F_2 = |d - d^*|. \tag{1}$$

Table 1. Problem specification data.

Parameter		Value
Support ring radius	r_s	0.25×10^{-3} m
Hemi-sphere radius	R	$2.8 - 3.2 \times 10^{-3}$ m
Shell thickness	h	$2 - 5 \times 10^{-6}$ m
Desired pressure	P^*	50×10^{-3} Pa
Desired displacement	d^*	24×10^{-6} m
Young's modulus	E	150×10^9 Pa
Poisson's ratio	ν	0.23

Table 2. Viable geometry parameters.

Number	$R \times 10^{-3}$ m	$h \times 10^{-6}$ m
1	3.140	2.234
2	3.127	2.242
3	3.119	2.244
4	3.098	2.229
5	3.097	2.228
6	3.093	2.223

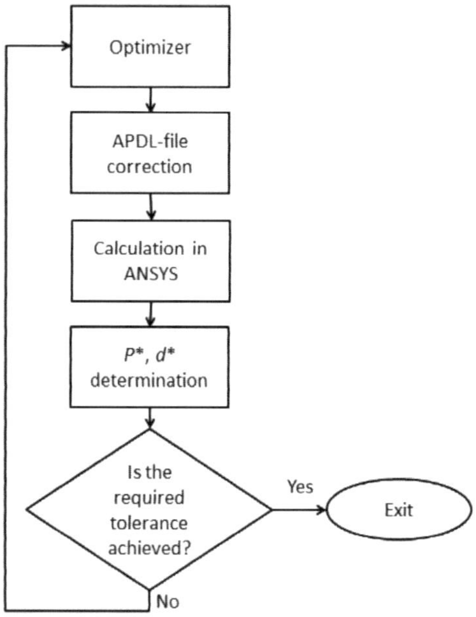

Figure 2. Flow chart of the solution in PSE/MACROS program.

An initial file is written in ANSYS Parametric Design Language (APDL), which performs the calculation within the ANSYS program. This file is held open and updated during the process of optimization. At each iteration, new values for the variables R (hemisphere radius) and h (shell thickness) are substituted into the APDL file, and the ANSYS analysis is repeated. The calculated values for the critical pressure, P, and the corresponding deflection, d, after the snap-through are extracted from the ANSYS results database using a subroutine written using Python language [22]. The Optimizer continues to vary R and h until the values of P and d match those of $P^*$and $d^*$to within the tolerance requirements.

3. Results obtained using a quasi-static modelling approach

In this optimization problem, 80 combinations of the two geometric parameters were considered during the solution of optimization problem.

A candidate solution is considered viable if it is near optimal in both objectives. Post-optimization, a further constraint might also be applied, $d \geq d^*$, to ensure that in the actuator the deflection is sufficient to ensure that a contact is closed. The 80 candidate solutions are projected onto the $P - d$ plane (Figure 3), allowing the user to choose the anticipated solution [23, 24, 25]. In this figure, there are six viable solutions, which are displayed in the in-set box [26]. The values of the variables R (hemisphere radius) and h (shell thickness) are displayed in Table 2, but it can be clearly seen that these are essentially identical, and the optimization solution is converged. The resulting shape, before and after snap-through is shown in Figure 4.

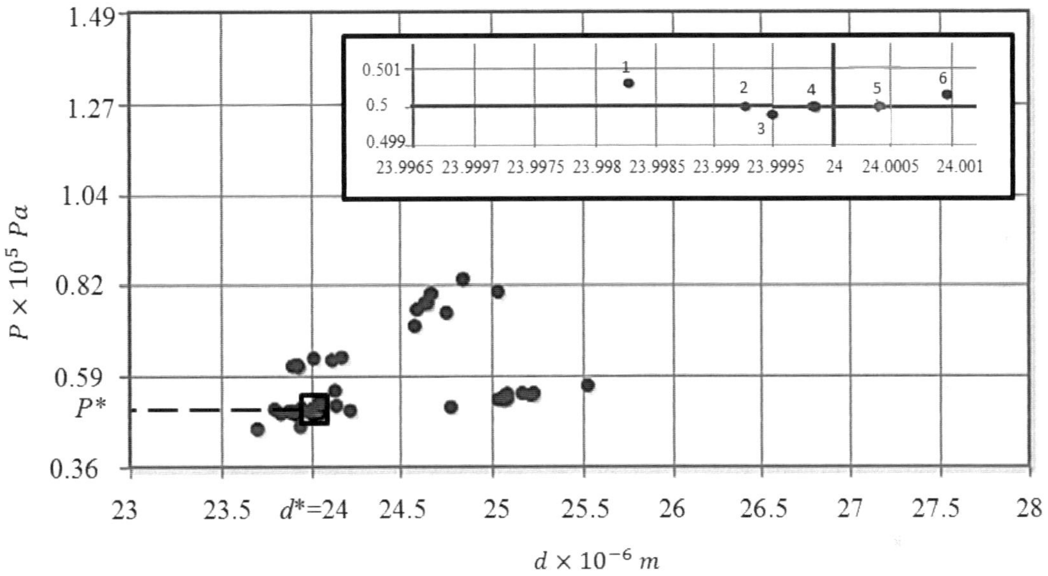

Figure 3. The $P - d$ plane.

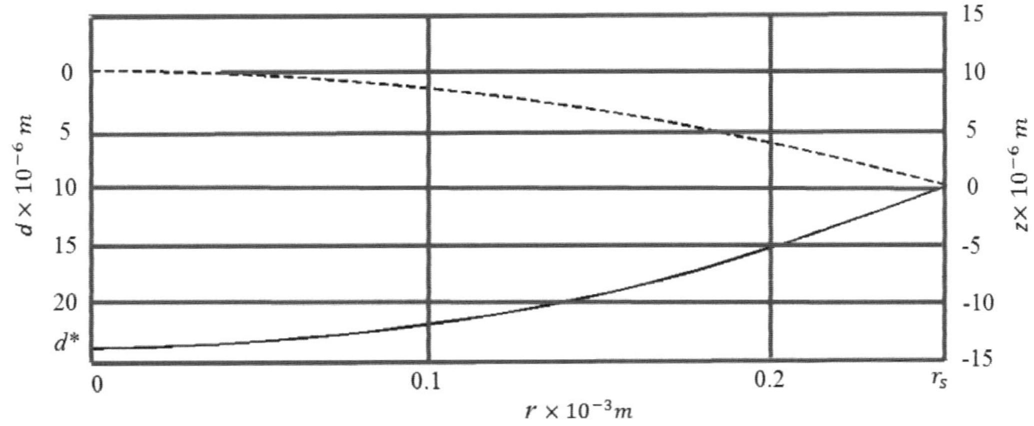

Figure 4. Initial shape (dashed) and shape after snap-through (solid) for optimised geometry.

4. Results where inertial effects were also considered

To determine whether it would be necessary to improve the Finite Element model of the actuator, the effect of inertia during the snap-through event was also considered. In this case it is again necessary to find viable geometrical parameters, and then to examine whether these differed significantly from those obtained assuming quasi-static deformation.

The procedure of solution for inertial loading in ANSYS is broadly similar, with the significant difference that in this case, the problem is solved incrementally in time, so it requires significantly greater computing power. This is because the explicit FEA formulation was used, which is "conditionally stable", hence the size of the time step increment is fixed by minimum element size. Because the FEA is time-consuming, the surrogate based optimization (SBO) method in PSE/MACROS program was used. Surrogate methods substitute an approximation for the computational model prediction, interpolated from results computed at a number of points in the problem domain. This is appropriate where the surrogate model gives a reasonable approximation of the model behaviour, and reduces the computational demand during the optimisation cycle [27].

For the solution of the optimization problem 17 combinations of geometric parameters were required, yielding optimal geometry values of $R = 2.94 \times 10^{-3}$ m and $h = 2.24 \times 10^{-6}$ m. The value for thickness is close to that obtained using the quasi-static modelling approach (less than 0.5% variance). The value for the hemispherical radius differs by approximately 5%, indicating that although inertia does influence the performance, it is quite a small effect, therefore it might not be necessary to take inertia into account during the initial phase of optimization. This could make the optimization process more efficient.

5. Discussion

The current research is at an early stage in development, and the focus here has been to develop an effective methodology for parameter optimization for structural components that display non-linear elastic large deflections. For this reason, the multi-objective optimization has been restricted to just two objectives, and the geometrical variables to just two parameters. Despite this simplification, the optimization problem has shown sufficient richness to demand further investigation.

PSE/Macros is a commercial off-the-shelf engineering design optimization package. As such, it is designed to be used by non-experts, to perform robustly, and to provide user-accessible results. It is neither under obligation to provide the user with details as to how it obtained those results, nor to enable the user to drive its operation in any particular way. The "black box" nature of the package restricts the user, such that it is difficult or impossible to deduce which optimization algorithms are being deployed, and impossible to impose a particular methodology or utilise a user-written algorithm.

For research purposes, it is necessary both to understand the processes of optimization, and to explore and prescribe particular actions and algorithms, firstly in order that the work might be validated by other researchers, and secondly to develop and improve the optimization algorithms for this particular class of optimization problem. For this to be feasible, the present authors recognize that future optimization work should be performed using either an open-source optimization package, or by developing in-house capability.

The refinement of the modelling capability to take into account the inertia effects of the actuator made some significant demands on the Finite Element computational time. To overcome this, the optimization step included surrogate modelling, and in doing so, the number of optimization steps required dropped from 80 to 17. Given the stochastic nature of optimization algorithms, it would be unreasonable to draw a firm conclusion from this single example, but this result would suggest that surrogate modelling could prove to be extremely valuable as the level of model geometry complexity is increased.

Optimization is currently a very busy research field, and there are very many new algorithms and techniques under development, for example [18, 23, 24, 25, 27, 28 29]. It is already known to concatenate algorithms; first to explore the parameter domain in order to determine how strongly variation of parameters influences the objective functions and to identify a region of the domain where a strong optimum is likely to be found, and subsequently to drive efficiently towards that optimum.

There are two particular areas of optimization research that would have particular relevance to the further development of micro-actuator design methods: the use of Pareto front mapping and penalty functions [24]. While a robust optimization technique might generate a number of viable design options, it would remain the duty of a human designer to make the final geometry selection. The present study has been a simplified paradigm of a much more complex problem: the choice between six options can be made on a case by case basis. Where the numbers of objectives is much greater, for example to include cost models for manufacture, raw material costs, materials utilisation, fatigue performance, *etc.*, Pareto front mapping could prove a useful tool in trading between the characteristics of different types of optima. In the measurement of the degree of an optimum there can sometimes be asymmetry. For example, in the present study the objective function for deflection was defined by minimising the difference between the deflection achieved by the candidate geometry, and the desired deflection. In point of fact, a slight over-deflection would probably be acceptable, while an under-deflection would mean that a contact would not be closed, and this would be entirely unacceptable. Penalty function methods can be applied to such asymmetric objectives, thereby screening out the candidates that do not meet practical requirements.

Analysis of the tolerance domain in the region of the "ideal" geometry parameters is also of significant importance. For a MEMS device, it is important that it operate without error, which means that every device must operate in the same way despite the inevitable geometric variability that would exist between individual devices. Should a high degree of tolerance be demanded, the cost of manufacturing would be inhibitive. Design optimization should therefore seek not for the absolute optimum, but for a robust optimum.

6. Conclusion

A technique for determining geometric parameters for a discrete action micro-actuator, to translate a prescribed critical pressure into a predetermined value of deflection has been described. The actuator was simplified to be a hemispherical shell, with a simply supported ring support of fixed diameter. The two geometry variables considered were the hemispherical radius, and the shell thickness. The mechanical analysis was performed using Finite Element Analysis. The mechanical model adopted was initially assumed to be quasi-static, but subsequently refined to consider inertia effects. The optimization process was performed using an off-the-shelf engineering design optimization package. The proposed methodology followed has been demonstrated to be effective, a number of opportunities for methods development have been identified, and these will form the basis for further more challenging design optimization studies.

References

[1] Ishihara H, Arai F, 1996, "Micro Mechatronics and Micro Actuators." *IEEE/ASME Transactions on Mechatronics* **1**(1) 68-79.

[2] Timoshenko SP, Gere GM, 2009, *Theory of elastic stability*, 2nd edition, Dover, New York 544.
Timoshenko SP, 1971, *Устойчивость стержней, пластин и оболочек. [Stability of rods, plates and shells]*, Nauka, Moscow [in Russian].

[3] Voitovich II, 1989, *Математические проблемы нелинейной теории пологих оболочек. [Mathematical problems of the nonlinear theory of shallow shells.]*, Nauka, Moscow [in Russian].

[4] Feodos'ev VI, 1969, "Осесимметричная эластика сферической оболочки." ["Axisymmetric elastic spherical shell."] *Prikladnaia matematika i mekhanika* [*Journal of Applied Mathematics and Mechanics*] **33**(2) 280-286 [in Russian].

[5] Grigoliuk EI, 1953, "Тонкие биметаллические оболочки и пластины." ["Bimetallic thin shells and plates."] *Inzhenernyi sbornik* [*Engineering collection*] **17** 69-120 [in Russian].

[6] Novozhilov VV, 2010, *Теория тонких оболочек [The theory of thin shells]*, Saint Petersburg

University, Saint Petersburg [in Russian].
[7] Reissner E, 1950, "On axisymmetrical deformations of thin shells of revolution." *Proc. Sympos. Appl. Math.* **3** 27-52.
[8] Popov EP, 1948, "Явление большого перескока в упругих системах и расчет пружинных контактных устройств." ["Snap-through in elastic systems and contact devices springs calculation."] *Inzhenernyi sbornik* [*Engineering collection*] **5** 62-92 [in Russian].
[9] Bich DH, Tung HV, 2011, "Non-linear axisymmetric response of functionally graded shallow spherical shells under uniform external pressure including temperature effects." *Int. J. Nonlinear Mechanics* **46**(9) 1195-1204.
[10] Li QS, Liu J, Tang J, 2003, "Buckling of shallow spherical shells including the effects of transverse shear deformation." *Int. J. Mechanical Sciences* **45**(9) 1519-1529.
[11] Petersen KE, 1982, "Silicon as a mechanical material." *Proc. IEEE* **70**(5) 420-457.
[12] Stefanescu DM, 2011, *Handbook of Force Tranducers: principles and components.* Springer, Berlin.
[13] Gavriushin SS, 1994, "Численное моделирование и анализ процессов нелинейного деформирования гибких оболочек." ["Computational modelling and analysis of the processes of nonlinear deformation of flexible shells."] *Izvestiia Akademii Nauk SSSR. Mekhanika tverdogo tela* [*Proceedings of the Academy of Sciences of the USSR. Mechanics of rigid body*] **1** 109-119 [in Russian].
[14] Valishvili NV, 1976, "Методы расчета оболочек вращения на ЭЦВМ." ["Methods for calculating the shells of revolution on a computer."] Mashinostroenie, Moscow [in Russian].
[15] Podkopaeva AS, 2012, "Анализ процесса деформирования микроактюатора. [Analysis of microactuator deformation process.]" *Machine Building* **S** 10-14 [in Russian].
[16] Zienkiewicz OC, 1977, *The Finite Element Method.* McGraw-Hill, New York.
[17] Lee JJ, Oh IK, Lee I, Rhiu JJ, 2003, "Non-linear static and dynamic instability of complete spherical shells using mixed finite element formulation." *Int. J. Nonlinear Mechanics* **38**(6) 923-934.
[18] Moaveni S, 2003, *Finite Element Analysis: Theory and Application with ANSYS.* Prentice Hall, New Jersey.
[19] Crisfield MA, 1981, "A fast incremental/iterative solution procedure that handles snap through." *Computer* & Structures **13**(1) 55-62.
[20] ANSYS Academic Research, 2012 Release 14.5, Help System, ANSYS Inc, Canonsburg.
[21] https://www.datadvance.net/ [last accessed October 2014].
[22] https://docs.python.org/3/tutorial/ [last accessed October 2014].
[23] Sobol' IM, Statnikov RB, 2006, *Выбор оптимальных параметров в задачах со многими критериями. [Choice of optimal parameters in problems with many criteria].* Drofa, Moscow [in Russian].
[24] Coello Coello CA, Lamont GB, van Veldhuizen DA, 2007, *Evolutionary algorithms for solving multi-objective problems*, Springer Science+Business Media, New York.
[25] Cortez P, 2014, *Modern Optimization with R*, Springer, Switzerland.
[26] Gavryushin SS, McMillan AJ, Podkopaeva AS, 2014, "Синтез микроактюатора дискретного действия по заданным функциональным параметрам." ["Synthesis of discrete microactuators from specified functional parameters."] *Machine Building* **1**(646) 55-60 [in Russian].
[27] Ong YS, Nair PB, Keane AJ, 2003, "Evolutionary optimization of computationally expensive problems via surrogate modelling." *Amer. Inst. Aero, Astron. J.* **41**(4) 687-696.
[28] McMillan AJ, Keane AJ, 1997, "Vibration isolation in a thin rectangular plate using a large number of optimally positioned point masses." *J. Sound Vibration* **202**(2) 219-234.
[29] Venkataraman P, 2002, *Applied Optimization with MATLAB Programming.* Wiley-Interscience, New York.

High sensitive materials in medical holographic microscopy

A Osanlou[1], E Snashall[2], O Osanlou[3], R Osanlou[4], E Mirlis[5], Lishen Shi[6] and H Bjelkhagen[7]

[1, 5, 6, 7] Glyndwr University, Mold Road, Wrexham, LL11 2AW, UK
[2] Whiston Hospital, Warrington Road, Prescot, L355DR, UK
[3, 4] University Hospital Aintree, Lower Lane, Liverpool, L97AL, UK

a.osanlou@glyndwr.ac.uk

Abstract. High sensitivity is defined in relation to the energy required to perform holographic recording. High sensitivity in silver halide materials is their main advantage over other similarly high resolving power holographic recording materials. This work reports progress on the development of silver halide based 'true colour holographic imaging', under a microscope. A thin layer of ultrafine grains of silver halide crystals of around 10 nm average diameter, dispersed in a colloid and coated on a substrate is used as the recording media. The significance of this method so far, is in its ability to produce 'true colour' three-dimensional images of specimen. The recordings have an appreciable depth, permitting the observer to scan through the image under a microscope, as one might with a real specimen sample. Current methods could perform 'True colour holographic imaging' directly under a microscope. The recording methodology has the potential for deeper complex and scattering media imaging, using very small pulses of appropriate laser wavelengths. The methodology, using novel nano-size panchromatic recording media consisting of dispersed fine nano grain crystals, could potentially revolutionise related medical imaging techniques. Future development of digital media will allow it to be utilized in this manner.

1. Introduction

Holographic microscopy in cellular imaging has emerged in recent years as a new tool for medical imaging application [1-16]. Current techniques allow imaging of very thin layers of scattering media. Stepping up these techniques to apply them for medical research is technologically challenging. The current generation of digital image sensors and spatial light modulators are limited in dealing with light diffraction [17].

This work reports on the development of 'True colour optical biopsy'. Rather than using digital imaging devices, a thin layer of ultrafine grain of silver halide crystals of 10-20nm average diameter, dispersed in a colloid and coated on a substrate is used as the recording media. The significance of this method so far, is in its ability to produce three-dimensional true colour images of specimen. They have an appreciable depth, permitting the observer to scan through the microscopic image, as one might with a real sample of cells, e.g. from biopsy. Current methods could 'True colour optical biopsy' the skin, the gastrointestinal tract and anywhere accessible with a direct scope imaging. This has the potential for deeper tissue holographic 'True colour optical biopsy' using very small pulses of appropriate laser wavelengths.

The hologram is a pattern of fringes formed by the interference of light scattered from the object and light which comes from the point reference source. The pattern contains total three-dimensional information of a scene, gathered in a single exposure. This provides a unique opportunity in microscopy in comparison to alternative microscopy methods, such as confocal microscopy or the traditional compound microscopy. Achieving this, requires considerations be given to; *(i)* the resolving ability of the recording media, to record the highest spatial frequencies of the interference patter created, and *(ii)* that the theoretical resolution is limited by diffraction. Recording a scene requires two light beams, of wavelength λ, to meet at the recording media at an angle θ. They form fringes of pitch $\lambda/(2n \sin(\theta/2))$. A small θ may result in noise in the form of zero order becoming difficult to separate from the true image. Phase shifting methods attempt to overcome this problem at a minimum expense [18, 22]. But, as θ approaches 180^0, higher resolution recording media is necessary to resolve the fine fringes produced. For a pixel of size X, the maximum wave number ζ_{max} that can be resolved is the inverse of twice the pixel size, *i.e.* $\zeta_{max} \sim 1/(2X)$ and $\zeta = \sin(\theta)/\lambda$. This limits the maximum angle between the reference and object beams. For example, for a device pixel size of 1.4μm, and a red beam of wavelength 633nm, the maximum angle would be $\theta_{max} = \sin^{-1}((633\times10^{-9})/(2\times1.4\times10^{-6})) \sim 13^0$. This limit can oppose the common diffraction resolution limit of the order of $z\lambda/a$, where z is the imaging distance and a is the aperture size. Reducing the distance z, between the sensor and the object, to improve resolution, results in fringe patterns too fine to be resolved by the imaging pixels. Much of the effort in digital holographic imaging is to compensate for related insufficiencies [17]. Most active technological efforts also include colour imaging, important in microscopy [21]. Holographic 'True colour optical biopsy' using suitable lasers and novel nano-size panchromatic recording media consisting of dispersed fine nano grain crystals, could potentially revolutionize surgical techniques. Future development of digital recording media will allow it to be utilized in this manner.

2. A flexible light laser engine and delivery system

2.1. *Industrial tasks*
Flexible, mobile, true colour holographic endoscopy requires a fibre delivered system with both suitable visible laser wavelengths combination and PM (Polarization Maintaining) transmission, preferably integrated in a single device. Gooch and Housego fibre optics laboratories (G&H) [30], with expertise to manufacture both RGB (red, green, and blue) sources and fused PM combiners were tasked with the development of a combination device with: *(i)* stable coherent light sources, operating at around 20mW in the red, green and blue regions at 658nm, 532nm and 452nm respectively; *(ii)* an output PER (Polarization Extinction Ratio) of greater than 15dB; and *(iii)* a transmission between 70% and 90% of each visible region. The final delivery system outputting RGB laser light through a PM Cable requires PM combining devices with SM inputs and a PM output. Figure 1(a) shows an example device fixed to an external substrate, and under green laser light operation. The choice of laser light and other variables were a compromise based on availability and suitability. Fabricating a low loss device that is both single-mode and polarization maintaining for all three RGB wavelengths proved challenging, but critical for a flexible holographic endoscopic imaging device. Figure 1(b), illustrates a simplified schematic of the device attached to a prospective endoscopic unit.

2.2. *A subsection*
As a direct result of this study, early innovations in fused fiber developments resulted in a G&H patent application US7653269B1. The methodology involves the use of a short stub of (non-PM) SM fiber which transmits blue light with low loss, as the "mixer region" in a fused coupler with PM inputs and outputs. The resulting world-first device with the combined output having a polarization extinction of 7dB was an excellent innovative achievement, however holographic 'True color optical biopsy' demands achieving greater than 15dB of polarization extinction at all three RGB wavelengths. In regard to progress with the required RGB PM lasers, so far three types of pigtailed single mode PM

laser have been developed: all commercial diodes pigtailed using in-house techniques and materials in an industry standard "14-pin butterfly" package, Figure 1(c), with integrated power supply as a "stand-alone" device. Figure 1(d) demonstrates a green, 532nm, PM pigtailed laser, using a frequency converted 808nm laser diode as the fundamental source, in a butterfly package. It has the ability to achieve stable fiber-coupling of up to 20mW and a polarization extinction of about 20dB. The blue 452nm pigtailed laser with greater than 20dB produced 5mW, but the red (660nm) pigtailed laser with a PER of greater than 20dB produced more than 40mW. The developments have been of industrial relevance in terms of original work relating to commercial success, particularly in the case of the 660nm pigtailed laser, leading the industrial partners to develop closely related devices for demands in the UK and globally.

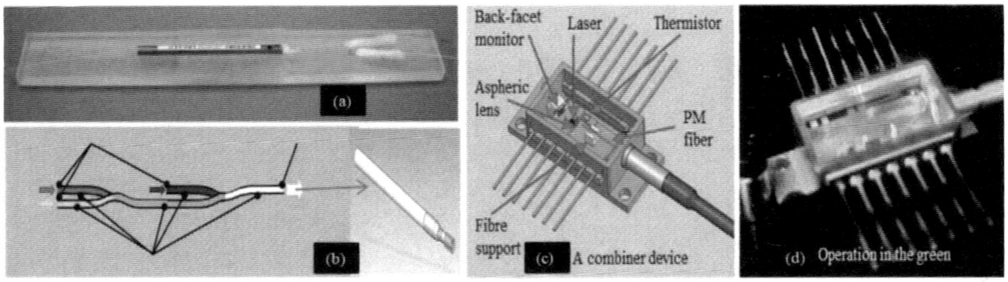

Figure 1. Images supplied courtesy of Gooch and Housego (G&H), (a) A PM combiner device with two SM inputs and a PM output, fixed to an external substrate, under green light operation, (b) a simplified schematic of the device attached to an endoscopic unit that should encapsulates a film holding unit, and (c, d) An industry standard "14-pin butterfly" package, operating in the green.

Figure 2. (a) The in-vitro recording setup, (b) schematic, (c) flexible beam delivery system for (d) delivery to the recording sample through a flat front silver mirror.

3. True colour recording

3.1. Novel recording media

Over the last few years, suitable low scatter panchromatic emulsions having a diffraction efficiency of around 95%, have been emerging. For example, the SilverCross material developed in house at Glyndŵr University [23] and used for true colour recordings reported in [24]. For commercially available recording emulsions, see for example [25-28]. For the recordings in this project, a newly developed version of the standard SilverCross was optimized in terms of large scale produce-ability and sensitivity at a recoding exposure of ~ 6mJcm^{-2} when machine coated. This compares with a recording exposure of ~ 4mJcm^{-2} for the same material when coated by hand, and ~ 1.7mJcm^{-2} for a representative commercial contender e.g. the Sphere-S material [25].

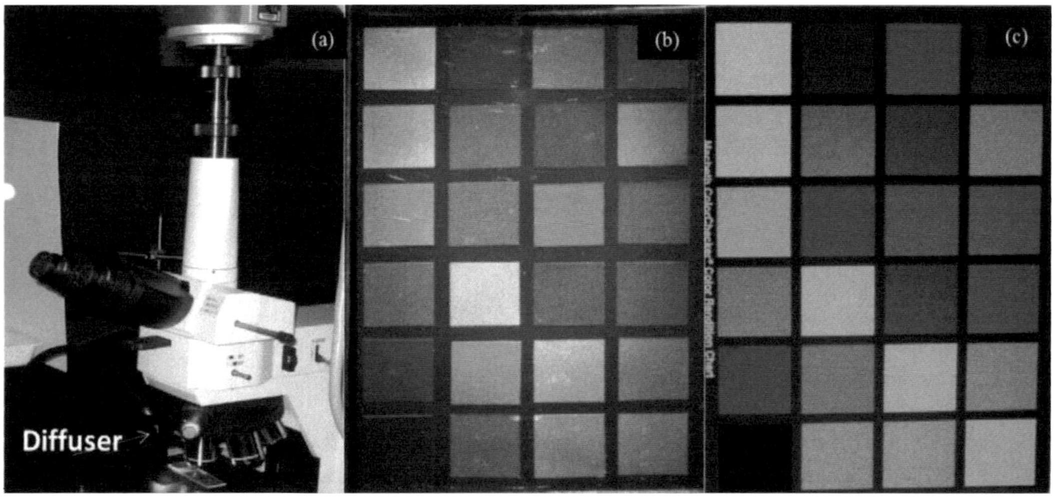

Figure 3. (a) "Non-coherent holographic reconstruction" under a microscope, of a (b) recorded hologram of the (c) Macbeth ColourChecker®.

3.2. In-vitro recording setup

The in-vitro setup is a simplified, but panchromatic version of an earlier monochrome in-vivo device, developed at Lake Forest College, dating back several years [31]. Utilizing dye spraying (or vital staining), the device demonstrated high contrast, in-vivo monochrome cellular imaging of the colon of an anesthetized dog. The images, however, lacked colour, particularly important to further aid physicians to distinguish unhealthy and dead tissue. This was the main clinical feedback of the trials. For the current recordings, while waiting for the development of a suitable specifically designed RGB fibre output device, the standard procedure involved panchromatic ultrahigh resolution in-vitro recording. The quality of the recorded colour holograms should be the same, independently of whether they are recorded in an open or fibre-optic holographic setup. The strategy, in the short run, would assist the development of the final holographic system, in demonstrating the best true colour three-dimensional reconstructed images, in terms of:

1. Ability to perform 'optical biopsy' of relatively large tissue areas in a single exposure
2. Not requiring extensive training to utilize the recording technique
3. Minimizing experimentation time by the clinical operator
4. Providing repeat examination opportunity e.g. by a specialist pathologist, viewing true colour, three-dimensional, large focal depth, sample image reconstructions.

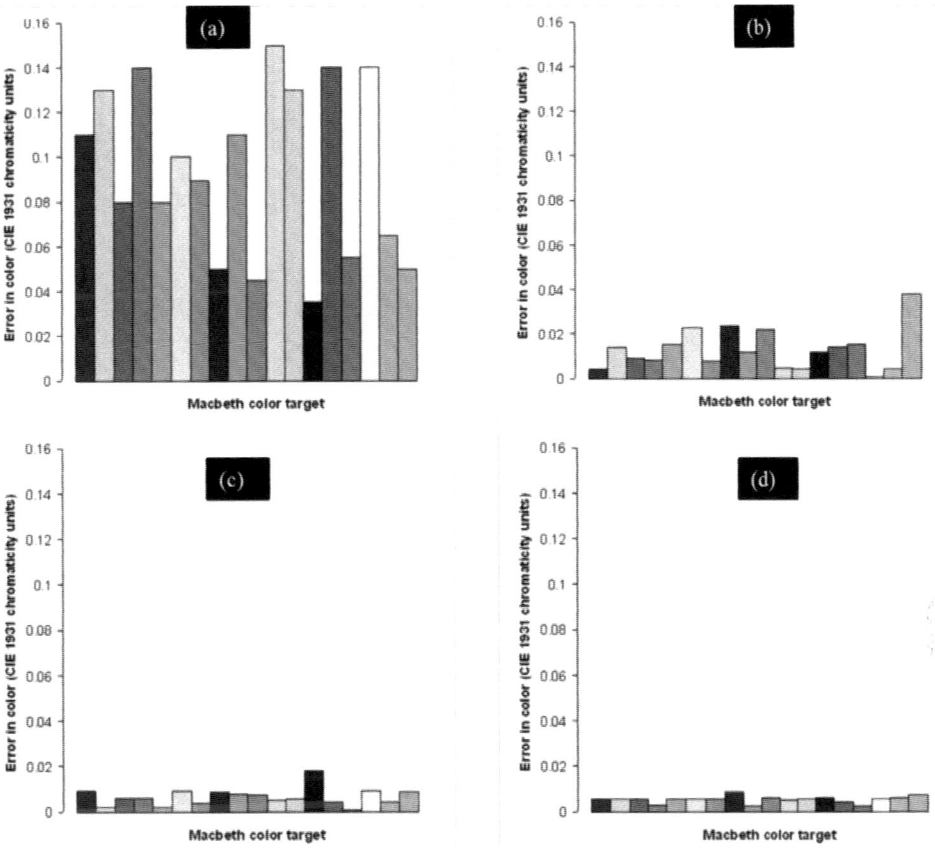

Figure 4. Colour error for each Macbeth colour for; (a) three wavelengths, (b) three optimum wavelengths, (c) four wavelengths and (d) five wavelengths.

To achieve these:

1. The setup, Figure 2(a-d), is of a 'white' laser Denisyuk [32] reflection type. The Denisyuk geometry provides maximum viewing angle and full parallax in both directions in the reconstructed images.
2. Keeping the recording media in close proximity to the tissue sample would also allow for minimum operator training and maximum recording area to be exposed to the 'white' wavelengths.
3. Because of the small object distance, the resolution obtained in the hologram is not compromised.
4. The setup contains no focusing element, providing three-dimensional reconstructions that are always 'in focus'.
5. The setup is hence designed to avoid focusing issues associated with small shakes and high magnification.
6. For the proposed recording setup, a commercially available single mode fibre was coupled to the following CW lasers to create the required 'white' beam: krypton ion laser (647 nm), frequency-doubled Nd:YAG (532 nm) and argon ion (476 nm).
7. The choice of visible wavelengths, and all other variables, including the fibre were a compromise based on closeness to optimum recording suitability, availability, and processing convenience [33].

4. Enhanced colour reproduction

Enhanced colour rendering or reproduction is possible if more than the minimum three primary RGB laser wavelengths are used [29]. In-house computer simulation, based on the tri-stimulus theory implications for the broad band light, and the narrow laser wavelengths, using the Macbeth ColourChecker® target and assuming a perfect recording media confirms this. The error calculations were based on the total average colour error in CIE 1931 chromaticity units for the optimal number of lasers, for different numbers of lasers, for all possible combinations between 700 and 400 nm [29].

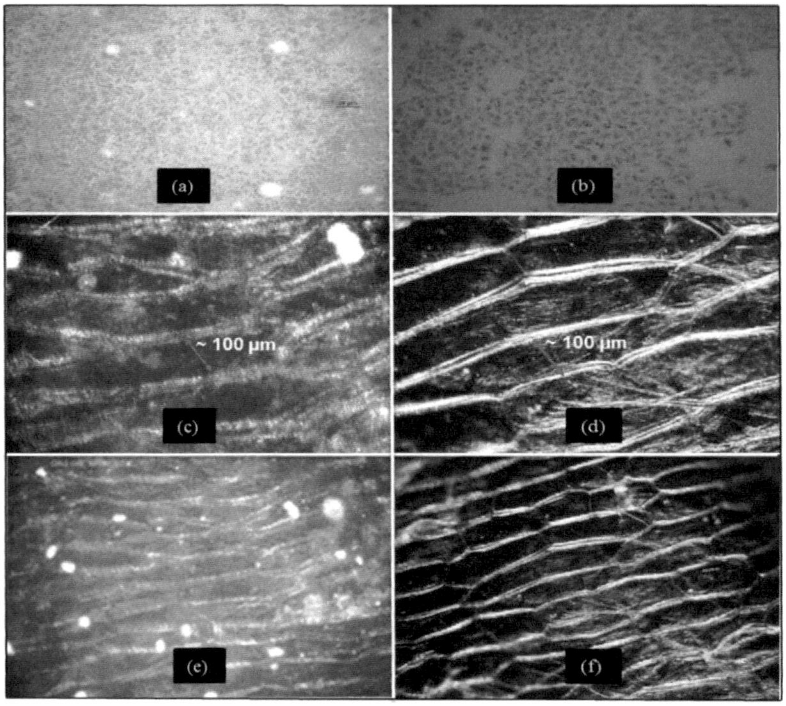

Figure 5. (a)'True colour optical biopsy' holographic reconstruction of (b) cultured human endothelial cells. (c) Hologram of (d) onion cells, both at twenty times magnification, also illustrated in (e) and (f) illustrated at ten times magnification.

A non-coherent broadband illumination source 'MEIJI Techno fibre optic (150W)' with a ground diffuser added to the beam line Figure 3(a) to enhance speckle reduction effect, was used for all holographic image reconstructions and image assessments. Figure 3(b) shows a non-coherent holographic image reconstruction of the Macbeth ColourChecker® Figure 3(c), utilized to check colour reproduction of holographic recordings of the in-vitro recording setup with 647, 532, 476nm wavelengths. The wavelengths were chosen because of availability and closeness to the optimum values defined by the simulation. Figure 4 (a) shows the simulation bar graph displaying the error for each Macbeth colour of the hologram: Figure 4 (b) shows the reduced error, if the optimal RGB wavelengths defined by the simulation 610, 545, 466nm were used. The reduced colour error for the optimum values of four laser wavelengths 620, 571, 518, 459nm is shown in Figure 4 (c), and for five lasers 452, 504, 549, 595, and 643 nm is shown in Figure 4 (d).

5. Colour in-vitro recording

For the holographic recording, Figure 5(a), of the human endothelial tissue, Figure 5(b), samples were cultured, stained, and supplied by expert medical clinicians, in a similar manner to that provided for medical investigations, *e.g.* by pathologists. Samples were then recorded holographically. Prior to recording endothelial cell samples, test holograms including that of the relatively large size onion cells, Figure 5(c-f), were recorded to assess and calibrate the setup. The medically approved reconstructed colour holographic images, demonstrating resolving cell nuclei, at different surface depth, Figure 6, were successfully investigated by medical experts, on the basis of "added quality information of medical relevance". Their main criteria were the ability to resolve the cell boundary and the nuclei inside the cells, and doing this at different depths of the surface of the three-dimensional space of the prepared human endothelial sample.

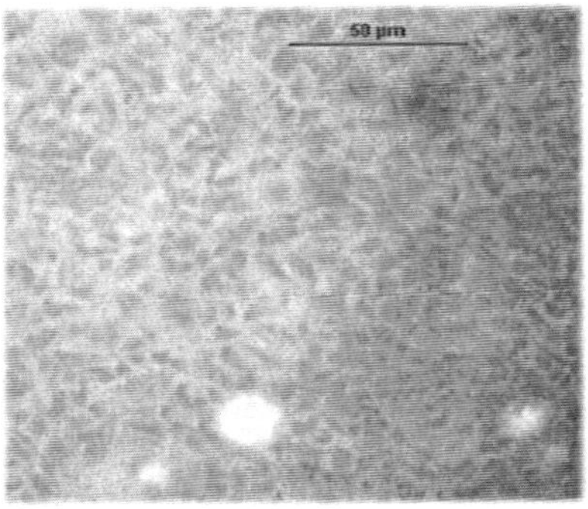

Figure 6. Development of medically approved color holograms from the human tissue demonstrating resolving cell nuclei, at different surface depth.

6. Discussion

Current methods could enable 'True colour optical biopsy' of the skin, the gastrointestinal tract and anywhere accessible with a direct scope imaging. This has the potential for deeper tissue 'True colour optical biopsy' using very small pulses of appropriate laser wavelengths. 'True colour optical biopsy' using novel panchromatic recording media consisting of dispersed fine nano grain crystals, could potentially revolutionize surgical techniques. Future development of digital recording media will allow it to be utilized in this manner.

References

[1] I. N. Papadopoulos, S. Farahi, C. Moser and D Psaltis, "Focused light delivery and all optical scanning from a multimode optical fiber using digital phase conjugation", Optical Fibers and Sensors for Medical Diagnostics and Treatment Applications XIII, edited by Israel Gannot, Proc. of SPIE **8576**, 857603-1 (2013).

[2] C. Haisch, "Optical tomography", Annu. Rev. Anal. Chem. **5**, 57–77 (2012)

[3] L. Kyeo Reh, K. Kyoohyun, J. Jaehwang, H. Ji Han, C. Sangyeon, L. Sangyun, C. Gyuyoung, J. YoungJu, P. Hyunjoo and P. YongKeun, "Quantitative Phase Imaging Techniques for the Study of Cell Pathophysiology: From Principles to Applications", Sensors**.** **13**, 4170-4191 (2013).

[4] S. R. Arridge, "Optical tomography in medical imaging", Inverse Probl. **15**, R41–R93 (1999).

[5] Y. Pu and D. Psaltis, "Seeing through turbidity with harmonic holography [Invited]", APPLIED OPTICS **52**, No. 4, 1559-128X/13/040567 (2013).

[6] Y. D. Chong and A. D. Stone, "Hidden black: Coherent enhancement of absorption in strongly scattering media", Phys. Rev. Lett. **107**, 163901 (2011).

[7] Z. Wang, K. Tangella, A. Balla, G. Popescu, "Tissue refractive index as marker of disease", J. Biomed. Opt. **16,** 10.1117/1.3656732 (2011).

[8] E. N. Leith and J. Upatniek, "Holographic imagery through diffusing media", J. Opt. Soc. Am. **56**, 523–523 (1966).

[9] E. Leith, H. Chen, Y. Chen, D. Dilworth, J. Lopez, R. Masri, J. Rudd, and J. Valdmanis, "Electronic holography and speckle methods for imaging through tissue using femtosecond

gated pulses", Applied Optics **30** (29):4204-4210 (1991).
[10] N. Abramson "Light-in-flight recording by holography", Optics Len. **3,** 121-123 (1978).
[11] C J Mann et al. "Dual Modality Live Cell Imaging with Multiple-Wavelength Digital Holography and Epi-Fluorescence", 3D Res**.** **2**, No. 01, 5 (2011).
[12] J Kühn, "Real-time dual-wavelength digital holographic microscopy with a single hologram acquisition", Optical Express **15**, No. 12, OSA 7242 (2007).
[13] Y Luo, "Spectrum resolved fluorescence imaging in multi-focal volume holographic microscopy", Proceedings of SPIE--the International Society for Optical Engineering **7891**, 78910C–78910C–5 (2011).
[14] T. Cižmár and K. Dholakia, "Shaping the light transmission through a multimode optical fibre: complex transformation analysis and applications in biophotonics", Opt. Express **19,** 18871–18884 (2011).
[15] S. Bianchi and & R.A. Di Leonardo, "Multi-mode fiber probe for holographic micromanipulation and microscopy", Lab Chip **12,** 635–639 (2012).
[16] Z. Yaqoob, D. Psaltis, M. S. Feld, and C. Yang, "Optical phase conjugation for turbidity suppression in biological samples", Nature Photon **2**, 110–115 (2008).
[17] D. E. Smalley, Q. Y. J. Smithwick, V. M. Bove, J. Barabas, S. Jolly, "Anisotropic leaky-mode modulator for holographic video displays" Nature **498**, 313–317 (2013).
[18] G. Brooker, N. Siegel, V. Wang, and J. Rosen, "Optimal resolution in Fresnel incoherent correlation holographic fluorescence microscopy", Optical Express v.**19**, No. 6, OSA, 5062 (2011).
[19] T. Tahara, T. Kakue, Y. Awatsuji, K. Nishio, S. Ura, T. Kubota, O. Matoba, "Parallel phase-shifting color digital holographic microscopy", 3D Research **1**, Issue, 1-6 (2010).
[20] P. Memmolo, M. Paturzo, A. Finizio, P. Ferraro and B. Javidi, "Synthesis and 3D display of multi-wavelengths digital holograms through adaptive transformation", Proc. SPIE **8738**, Three-Dimensional Imaging, Visualization, and Display, 87380N (2013).
[21] L. Yu and M. Kim "Full-color three-dimensional microscopy by wide-field optical coherence tomography", Optics Express **12**, Issue 26, 6632-6641 (2004).
[22] D. P. Kelly, B. M. Hennelly, N. Pandey, D. P. Kelly, B. M. Hennelly, N. Pandey, "Resolution limits in practical digital holographic systems", Optical Engineering **48**, 9_095801 (2009).
[23] A Glyndŵr University European CRAFT project COOP-CT-2004-5901, Centre for Photonics Studies, School of Engineering, Institute of Arts, Science and Technology, OpTIC, St Asaph Business Park, St Asaph, North Wales, LL17 0JD UK.
[24] H.I. Bjelkhagen and A. Osanlou, "Color holography for museums: bringing the artifacts back to the people", Proc. SPIE: Practical Holography XXV: Materials and Applications **7957**, 0B-1-7 (2011).
[25] Sphera–S Ltd, Russia (01,August, 2013); www.geola.com
[26] Ultimate – Holography France (01,August, 2013); www.ultimate-holography.com
[27] Colourholographic Ltd UK (01,August, 2013); www.colourholographic.com
[28] Bayer MaterialScience AG, Germany (01,August, 2013); www.bayermaterialscience.com
[29] E. Mirlis, H. I. Bjelkhagen and M. Turner, "Selection of optimum wavelength for holography recording", Practical Holography XIX: Materials and Applications, Proc SPIE **5742**, 113-118 (2005).
[30] Gooch and Housego PLC, Broomhill Way, Torquay TQ2 7QL UK (01,August,2013); http://www.goochandhousego.com
[31] M. D. Friedman, PhD thesis, Northwestern University, USA (1989).
[32] M.A. Klug, "Display applications of large scale digital holography", Proc. SPIE: Holography: A Tribute to Yuri Denisyuk and Emmett Leith **4737**,142-149 (2002).
[33] H.I. Bjelkhagen, " Silver halide recording materials for holography and their processing", Springer Series in Optical Sciences **66** , Heidelberg, New York: Springer (1993).

Surface hardening of optic materials by deposition of diamond like carbon coatings from separated plasma of arc discharge

AS Osipkov[1], VM Bashkov[1,2], AO Belyaeva[1,2] R Stepanov[1], YM Mironov[1], AL Galinovsky[1]

[1] Bauman Moscow State Technical University, 2-ya Baumanskaya str. 5, Moscow, 105005, Russia
[2] New Plasma Technologies LLC, B. Semenovskaya str. 49, off 202-205, Moscow, 107023, Russia

osipkov@bmstu.ru

Abstract. This article considers the issue of strengthening of optic materials used in the IR spectrum by deposition of diamond like carbon coatings from separated plasma arc discharge. The report shows results of tests of bare and strengthened optical materials such as BaF2, MgF2, Si, Ge, including the testing of their strength and spectral characteristics. Results for the determination of optical constants for the DLC coatings deposited on substrates of Ge and Si, by using separated plasma, are also presented. Investigations showed that surface hardening of optical materials operable in the IR range, by the deposition of diamond like carbon coating onto their surface, according to this technology, considerably improves operational properties and preserves or improves their optic properties.

1. Introduction

Optoelectronic systems, operating in IR range are widely spread in different fields of contemporary science and engineering. Such systems enable monitoring of various processes and objects at night and in low visibility conditions.

Enhancement of operational characteristics and efficiency improvement of such systems are inseparably linked with use of specific optical materials, such as: special kinds of glass (oxygen-free glass, and optical ceramics based on polycrystalline ZnSe ZnS, and used in the manufacture of optical components for IR devices, substrates and filters), optical crystals (CaF_2, MgF_2, BaF_2, KBr, NaCl, used to create lenses for infrared lenses, prisms, and window input devices for spectral measurements), and semiconductor materials (*e.g.* Si and Ge, which are used to create optical elements for thermal imaging systems operating in the range of and 3-5 and 8-12 microns).

Optical surfaces of the aforementioned materials are often soft and hygroscopic; as a consequence they can be easily damaged during operation through the effect of atmospheric moisture, dust and various mechanical impacts. Protecting such surfaces assumes strategic and actual importance.

Research into optical coatings is carried out in numerous institutions. Among Russian institutions engaged in such research one should note the scientific and industrial corporation *Vavilov State Optical Institute, NII OEP PLC,* the research institute *Radioelectronics and Laser Technology,* and *Bauman Moscow State Technical University*; and outside of Russia, centres such as *Nano-Care AG (USA), Corning (USA),* and *Izovac Ltd. (Republic of Belarus)*, however the results of studies achieved are uncoordinated and cannot currently be used to create a nomenclature of optical materials.

Therefore this study is aimed at the development and introduction of the technology to increase abrasive surface solidity and micro-hardness of optic materials from various classes – "soft" materials for UV and IR range, semiconductors, metals, polarized films and intermediate layers of switches of laser radiation. The basic methods of strengthening optic materials can be divided into two groups: creation of compressive stresses in the surface layers of optic materials (*e.g.* hardening, ion exchange, surface crystallisation) and local surface strengthening (*e.g.* thermal and thermochemical polishing, deposition of strengthening coatings chemically connected with the surface). The last method is one of the most efficient ways of strengthening, but currently carbon-based coatings are of particular interest [1].

The aim of this study is to evaluate the potential for the protection of optical elements operating in the IR range by the deposition of a diamond like carbon (DLC) coating on the surface by the electro-arc technique with laser ignition of carbon plasma and separation of plasma flow [2].

To achieve this goal the following scientific technical tasks were solved:

1. Deposition technology for DLC coating of non-organic "soft" (BaF2, MgF2) and semiconductor (Si,Ge) materials in the IR Range was developed.
2. Production of prototypes materials with DLC coatings deposited onto their surfaces.
3. Development of techniques for investigating the performance and optical properties of these prototypes.

2. Potential of hardening coatings to provide anti-reflection optical performance

The aim of this study is to not only increase the hardening characteristics of the optical materials, but also to preserve or improve their optical properties, by reducing the coefficient of reflection from their surface. Therefore the possibility of increasing transmission by means of enhanced anti-reflection performance as a result of the deposition of hardening coating requires special attention.

The refractive indices of the optical materials requiring hardening mentioned above, vary over a wide range of values from 1.4 to 4. It is known that for optimal single-layer anti-reflection, the refractive index of an applied coating has to meet the following condition

$$n_f \approx \sqrt{n_s} \tag{1}$$

where n_f is the refractive index of a film and n_s is the refractive index of a substrate material.

Thus, to increase the transmission of substrates made from highly-refractive materials (Si, Ge), which have refractive indices in the range from 3.5 to 4, it is necessary to deposit hardening films with refractive index about 1.7 to 2. For substrates with refractive indices in the range from 1.4 to 1.6 (BaF_2, MgF_2, CaF_2, KBr), coatings should have refractive indices of approximately 1.15 to 1.3.

3. DLC deposition technology and research methods

Diamond like carbon coatings are a class of coatings consisting of sp^3- and sp^2-hybridised carbon atoms. As a result of their unique chemical and physico-mechanical properties these coatings are widely used in various engineering fields from optics to heavy mechanical engineering and medicine. These properties include high hardness (5000-10000 Hv), low friction coefficient (0.15-0.08), high wear resistance and chemical inertness.

These coatings are usually produced by chemical vacuum deposition using HF plasma (the most common method), or magnetron sputtering of a graphite target and pulse laser sputtering (ablation). The properties of DLC coatings depend on the methods and conditions of deposition, because these give a different hydrogen content in the coating material and thus a different ratio of the sp^2- to sp^3-hybridised carbon atoms.

In this work we investigated coatings deposited by the electro-arc technique, with laser initiation of an arc and separation of a plasma flow. This was performed on the rig East 01 (New Plasma

Technologies LLC, Russia, Figure 1). The technological process is based on the separated arc evaporation of carbon from rotating cylindrical graphite target (cathode).

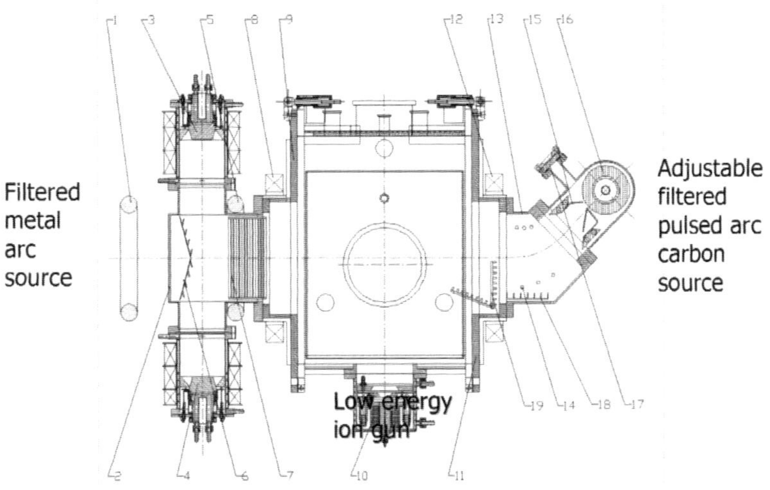

Figure 1. Schematic of a rig for diamond like carbon coating deposition.

1,5: solenoid; 2: camera filter; 3,4: titanium target; 6,7: trap particles; 8,12: solenoids for plasma confinement (Helmholtz coils); 9,11: water cooling; 10: ion gun; 13: camera filter; 14: carbon filters; 15: anode; 16: graphite cathode; 17: window for the laser; 18, 19: filtration system

When developing this technology, the nano-composite hardening was created using an approach based on the formation of various combinations of the following three effects: multiphase hardening, grain size effects, and the effect of suppression of grain boundary shift by the formation of strong contacts.

By varying the parameters marked by *, the following technological modes were defined: power of laser radiation: 20 mJ; frequency of laser radiation: 20 Hz; cathode voltage*: 250 V; hold-in solenoid current*: 2 A; supply* of working gas (Ar): 15 cm^3/h; deposition time*: 60 min; chamber temperature*: 34 °C; rotating velocity* of the turntable on which the samples were fixed: 3 rpm; carbon source voltage*: 200 V; and chamber pressure*: 10^{-5} Mbar.

Measurements of thickness, roughness, micro-hardness and optical characteristics of deposited layers and tests of their surface hardness were performed for applied DLC coatings. Moreover, the similarity of coatings to diamond was estimated, *i.e.* the ratio of sp^2-and the sp^3-hybridised carbon atoms in the coating. Roughness and thickness of coatings were measured by means of atomic-force microscopy using a rig Ntegra Spectra (NT-MDT, Russia). For thickness measurement of a substrate, a step on its surface was created by masking a part of the sample before coating deposition.

Surface hardness measurements of the samples before and after the deposition of hardening layers were performed on Nanoindentor Hysitron TI750 UBI with Oliver-Farrah's (ISO 14577) using a Berkovich-type indentor, which is a tri-hedral diamond pyramid with an angle at the vertex of about 142°. Surface hardening was estimated according to OST 1901-95 where it was defined quantitatively by its ability to resist surface abrasion under load. The abrasion resistance was characterised quantitatively by examining the scratch formed through a magnifying glass. All coatings were sorted into five groups on the basis of mechanical hardness.

The ratio of sp^2- and sp^3-hybridised carbon atoms in the coating was estimated by means of Raman spectroscopy using the rig Ntegra Spectra. The carbon phase of the coating was observed in the range from 1100 to 1700 cm^{-1}. The phase decomposition was carried out using the technique described in [3]. The D-peak was prescribed by Lorentzian, G-peak – by Breyt-Wigner-Fano function.

Spectral characteristics of the DLC coatings deposited were measured using a PerkinElmer Lambda 950 spectrometer, in the range from 0.2 to 3 microns, and an IR- Fourier spectrometer VERTEX 70 V, from 3 to 20 microns.

To define the optical characteristics of the films and thicknesses, according to the results of multispectral transmission measurements of a film-substrate system, a quasi-decision selection method was undertaken on the basis of the minimisation of the residual function [4]

$$F = \sum_{i=1}^{N} \left[T(\lambda_i)^2{}_{\text{exp}} - T(\lambda_i)^2{}_{\text{mod}}(n_1, k_1, d_1 \right]^2 \qquad (2)$$

where N is the number of radiation wavelengths at which measurements are made; $T(\lambda_i)^2{}_{\text{exp}}$ and $T(\lambda_i)^2{}_{\text{mod}}$ are the experimentally measured and modelled (respectively) values of the transmission coefficients at wavelength λ_i, and n_1, k_1 and d_1 are the refraction coefficients, absorption and film thickness respectively.

For the model of the film transmission on the substrate, a variant of the matrix method [5] was used. This allows the user to take into account the possible deviations from an ideal case, such as inhomogeneity of layers and roughness of transitions between them, and the existence of "thick" layers.

The minimisation of equation (2) is a complex optimisation problem, as the residual function has a large number of local minima, and cannot always be defined in all areas of a search. The number of parameters can reach 10 and more, and include: film thickness, coefficients of model dependence, $n_1(\lambda)$ (*e.g.* coefficients of Sellmeier equation), or coefficients of model dependence, $k_1(\lambda)$.

Nowadays there are a number of software products for the determination of film parameters using multi-spectral measurements, however the majority of such programs are problem specific and do not allow flexible solving of numerous problems arising while determining of optical characteristics of diamond like nano-films. A special software program [4] providing measurements of thickness and optical characteristics of diamond like carbon nano-films was developed in the visual programming environment LabVIEW 2011, in which two optimisation algorithms are implemented using the method of consecutive square programming [6], and the method of differential evolution [7].

4. Results

The Raman spectroscopy analysis of the chemical composition of the deposited carbon layers showed (Figure 2) that the ratio of sp^3- and the sp^2-hybridised carbon atoms varies from 40 to 65% on the substrates oriented directly perpendicular to the plasma flow. On the parallel substrates, this indicator is much lower (from 17 to 35%) which might be due to different conditions of carbon film growth.

Analysis of the mechanical characteristics of the surface layer (Table 1) of the optical materials considered, deposited on various types of substrates, showed that DLC coating deposition leads to a surface hardness increase from 18% for Ge, to 250% for a substrate made from soft BaF2, accompanied by an increase in elastic modulus of the hardened surface of between 10 and 36%. Thus the tests of mechanical hardness of samples of Ge and Si showed that DLC coating deposition provides a surface hardening corresponding to "0" hardness class according to OST 1901-95.

An increase of surface roughness was observed on substrates from soft materials. This is might be due to surface layer erosion of such materials under the influence of ion beams. The matter demands further study, and the optimisation of the technological process for these materials.

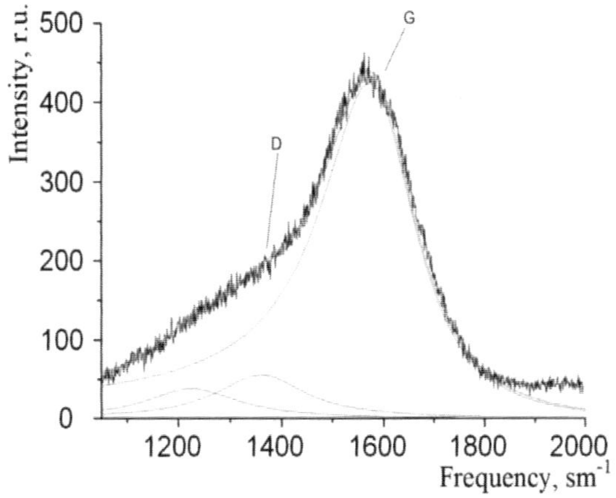

Figure 2. Typical Raman spectrum of DLC-coating and its decomposition (Ge substrate, sp^3 – 48%)

Table 1. Mechanical characteristics of DLC coatings

Substrate material	Coating thickness, nm	Substrate material hardness, GPa	DLC-coating hardness, GPa	Substrate surface modulus of elasticity, GPa	Hardened surface modulus of elasticity, GPa
BaF_2	72	1.7	4.3	66.03	90.4
MgF_2	71	7.2	14.7	136	162.4
Si	65	12.10	20.15	116.5	130.7
Ge	101	12.10	14.36	-	-

Measured spectral transmission characteristics of the studied on the samples are presented in Figures 3 and 4. On the basis of the analysis of the received spectral characteristics one can conclude that after hardening, DLC coating deposition preserves the spectral range of transmission of the prototypes in the IR range for BaF_2 and MgF_2, and also some anti-reflection for Ge and Si samples is observed.

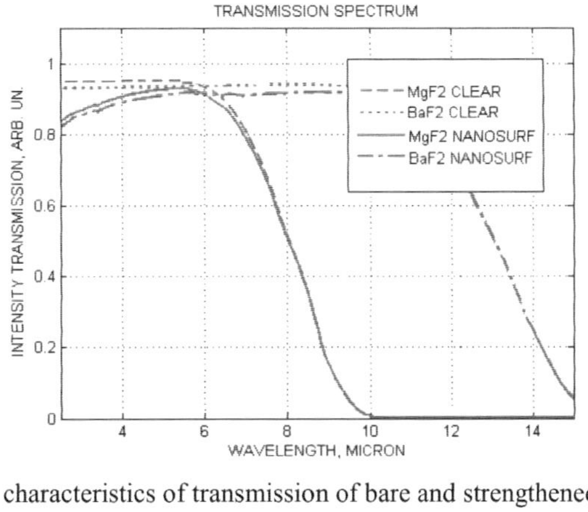

Figure 3. Spectral characteristics of transmission of bare and strengthened MgF_2 and BaF_2

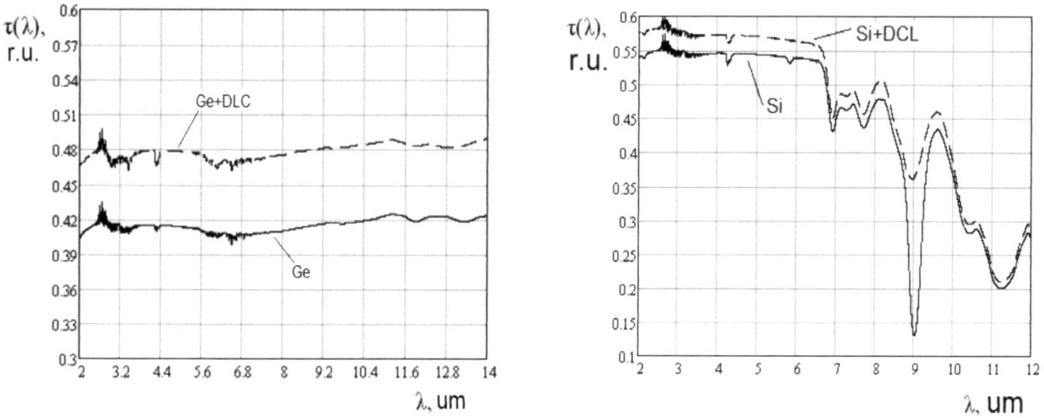

(a) (b)

Figure 4. Results of measurement of spectral transmission characteristics of bare and gardened surfaces of (a) Ge, and (b) Si.

Results of determination of optical constants for DLC on substrates of Ge and Si are presented in Figure 5. Distinction of optical coefficients for DLC coatings on substrates of Ge and Si may be due to different conditions of carbon layer formation while deposition, and as a consequence, the different structure of a formed DLC film.

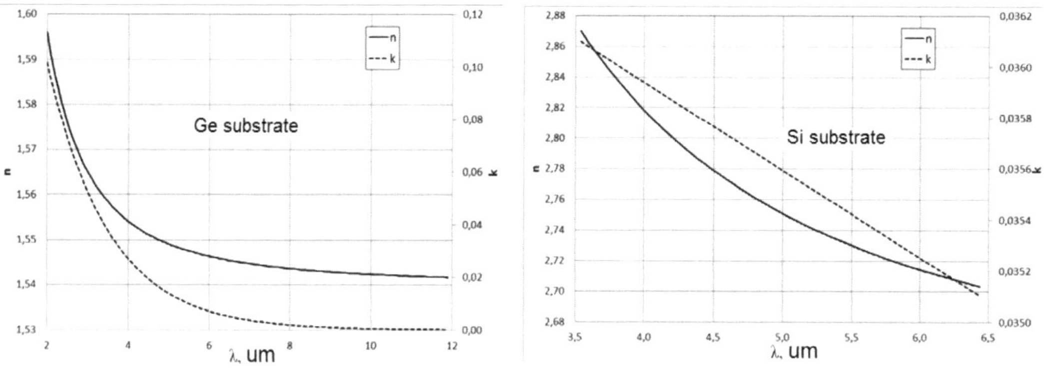

(a) (b)

Figure 5. Optical characteristics (refractive index n and absorption coefficient k) for DLC coating on substrates of (a) Ge and (b) Si.

5. Hardening of anti-reflective coatings

Surface hardening of "soft" optical materials while preserving their optical properties can be considered as an important practical result, capable in increasing considerably the operational properties of these materials. At the same time semiconductor materials, as a rule, also require anti-reflection charateristics. During execution of this research, samples of substrates were made using the semiconductor materials Ge and Si with thin layer Al2O3-Ge anti-reflection coatings. Hardening

IOP Conf. Series: Materials Science and Engineering **74** (2015) 012013 doi:10.1088/1757-899X/74/1/012013

layers of DLC were subsequently deposited onto the surfaces of the aforementioned substrates. The measured transmission characteristics are given in Figure 6. Results of micro-hardness measurements of the hardened anti-reflective coatings are given in Table 2.

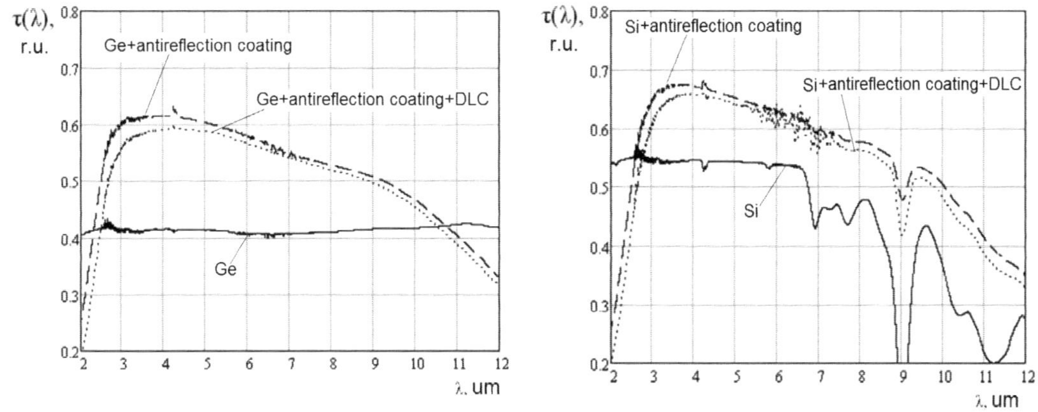

Figure 6. Optical characteristics (refractive and absorbtion coefficients) of DLC coating on substrates of (a) Ge, and (b) Si.

Table 2. Results of micro-hardness measurements of Si and Ge samples covered with hardened anti-reflection coating.

Substrate	Hardness, GPa		
	Sample without coating	Sample with antireflection coating	Sample with antireflection and DLC coating
Si	11. 1	9.8	15.3
Ge	12.5	4.1	16.8

The data analysis presented in Figure 6 and in Table 2 indicates that, with development, the DLC coating technology has the potential to preserve the desired IR transmission characteristics for optical details of the semiconductor materials Ge and Si and with anti-reflective coatings, and to improve their mechanical strength properties.

6. Conclusion

Experimental research showed that surface hardening of optical materials operated in IR range, by deposition of diamond like carbon coatings, using our developed technology of electro-arc deposition with separation of plasma flow and laser ignition of an arc, considerably improves the operational properties of these materials and preserves or improves their optical properties. The developed DLC deposition technology, including anti-reflective coatings, should be of great practical interest for developers of optical systems for modern thermo-vision devices working in conditions of aggressive mechanical impacts on optical elements.

References

[1] Kamanina N, Vasilev P, Studenov B, 2008, "Nanotechnology applications in optics: a possible increase transparency and increase the mechanical strength of the surface material UV and IR spectral ranges." *J. Opt. Technol.* **75**(12) 57-60.

[2] EP1884978B1 "Process for the coating of substrates with diamond-like carbon layers CREEPSERVICE" SARL, Bole, CH (Chizik A, Mikhailov A) 06016249.2 03.08.2006 Publ. 19.10.2011 p 92.

[3] Ferrari AC, Robertson J, 2000, "Interpretation of Raman spectra of disordered and amorphous carbon." *Phys. Rev. B* **61** 14095-107.

[4] Fedotov Y, Stepanov R, 2013, "Software for determining the optical characteristics of ultrathin films Engineering and scientific applications based on technologies." *National Instruments – 2013: Proc. of the XII Int. scientific and practical conf.* 147-149 (Moscow: DMK-Press).

[5] Katsidis C, Siapkas DI, 2002, "General transfer-matrix method for optical multilayer systems with coherent, partially coherent, and incoherent interference." *Applied optics* **41** 3978–87.

[6] Bonnans JF et al., 2006, *Numerical optimization: Theoretical and practical aspects*. Springer-Verlag, Berlin.

[7] Price K, Storn RM, Lampinen JA, 2005, *Differential Evolution: A Practical Approach to Global Optimization*. Springer–Verlag, Berlin.

Aluminium Sheet Metal Forming at Low Temperatures

R Schneider[1,2], B Heine[2], R J Grant[1] and Z Zouaoui[1]

[1] Glyndŵr University, Mold Road, Wrexham, LL11 2AW, UK
[2] Aalen University, Beethovenstr. 1, Aalen, 73430, DE

robert.schneider.wrexham@hotmail.com

Abstract. Low-temperature forming technology offers a new potential for forming operations of aluminium wrought alloys which show a limited formability at ambient temperatures. This paper indicates the mechanical behaviour of the commercial aluminium alloys EN AW-5182 and EN AW-6016 at low temperatures. Stress-strain relationships at different temperatures were investigated through tensile testing experiments. Flow curves were extrapolated using an adapted mathematical constitutive relationship of flow stress and strain. A device which allows cupping tests at sub-zero temperatures was specially designed and a limiting dome height was determined.

1. Introduction

Statutory provisions of CO_2 emissions require constructive measures regarding the weight reduction of a car body. An option to reduce the nominal weight of automotive structures is to use light-weight materials, such as aluminium alloys, instead of steel [1, 2]. Commonly used aluminium alloys in the sheet metal processing industry are EN AW-5182 for interior panels and EN AW-6016 for outer skin applications. It is known that the forming behaviour for certain aluminium alloys is limited when formed at ambient temperatures [3, 4].

An option to enhance the forming behaviour of currently used aluminium alloys would be a thermal heat treatment application [5], but thermally induced forming operations may influence the mechanical behaviour of such alloys. Thus, forming aluminium alloys below room temperature would be an approach, where strain hardening losses and precipitation processes are not an issue. In this regard there is no danger of a reduction in strength through the mechanism of recovery, recrystallization and aging which is in most cases induced by forming at higher temperatures.

Based on limited published data [6-8] where the idea of improving the formability of aluminium and its alloys by a low temperature treatment is mentioned, the potential and the feasibility of such a technique is still questionable, however, in the current research, stretch forming investigations have been performed to show the applicability of the low temperature treatment for multi-axial stress states.

2. Experimental

2.1. *Materials*

For this study aluminium alloy sheets with the designation EN AW-5182 H111 and EN AW-6016 T4, showing a thickness of t = 1.15 mm, were used. Table 1 illustrates the composition specification of tested aluminium alloys.

Table 1. Chemical composition of tested aluminium alloys (wt.-%)

	Si	Fe	Cu	Mn	Mg	Cr	Zn	Ti
AA5182	0.20	0.35	0.15	0.35	4.40	0.10	0.25	0.10
AA6016	1.20	0.40	0.15	0.15	0.40	0.10	0.15	0.15

2.2. Laboratory

The experiments were performed using a tensile testing machine from Zwick with a 100 KN loading cell. For data processing the software package testXpert V 11.0 was used. The set-up shown was implemented into the tensile testing machine where forces and displacements could be monitored.

<u>Tensile tests:</u>

Flat sheet specimens were CNC machined showing a test section width of 12.5 mm in accordance with DIN EN ISO 6892-1 (A_{50}) [9]. Attention was given towards the rolling direction of the sheet metal which was perpendicular to the applied load. The length of the 50mm wide clamp sections of the specimens were designed to fit with the requirements of the cryogenic set-up. A specially designed cryogenic dewar, Figure 1, was used to obtain and maintain the sub-zero temperatures required in these tests. A variation of testing temperatures was reached by way of a *medium-pipe*, which was used as a vessel carrying isothermal media, and was cooled by liquid nitrogen (LN_2). The final temperature was reached in the cryogenic dewar, where the specimen remained for around 5 minutes, before the test started. Furthermore, homogeneity of the temperature within the medium-pipe was accomplished by injecting inert gas. In case of a sudden temperature drop the chamber included an inlet for compressed air, which ensured a replacement of the cold environment between medium-pipe and cryogenic dewar. The tensile tests were conducted at five different temperatures of 25, -50, -100, -150, and -196 °C with a forming velocity of v_f = 200 mm/min (maximum velocity of the tensile testing machine) resulting in a strain rate of $\dot{\varepsilon_f}(v_f) = 6.60 \times 10^{-2}$ s^{-1}.

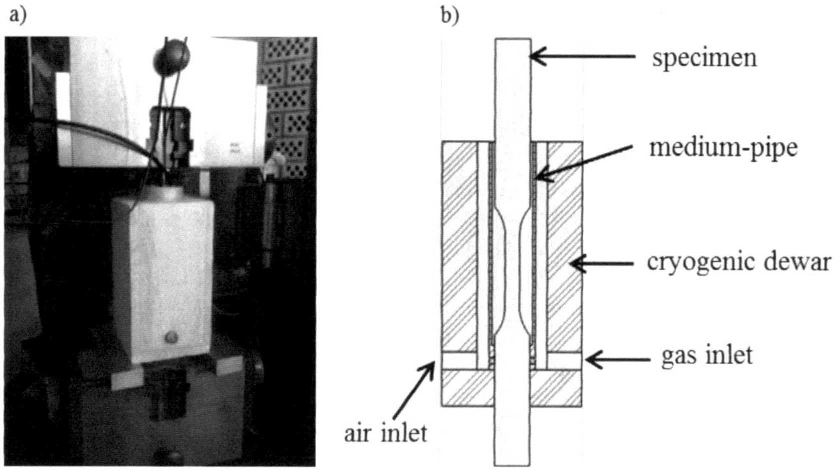

Figure 1. a) Tensile test equipment; b) cross-sectional view of tensile test set-up.

<u>Limiting dome height (LDH):</u>

A stretch forming apparatus capable of operating at very low temperatures was designed for evaluating the LDH (Fig. 2). LN_2 was used for cooling the apparatus to sub-zero temperatures. The temperature was monitored via thermocouples placed at the punch, blank holder, die, and blank. A cross-sectional view of the geometry used in the tests is shown in Fig. 2b which is based on [10]. Pre-investigations showed that by using graphite based anti-friction agents a moderate lubrication of the system is still achieved at low temperatures. Thus, both sides of the quadratic blank (100 × 100 mm) were wetted using graphite suspension before forming. Afterwards, the blanks were cooled to the desired forming temperature inside the testing apparatus via conductivity. Constraining the flow of the blank material into the die was achieved by tightening six carriage bolts positioned in a circle, Figure 2a. As a stop criterion for the test a punch load-drop of 50 N was taken as an indication of necking in the specimen. The punch velocity was set to be v_p = 50 mm/min. For statistics five experiments were carried out at each temperature level.

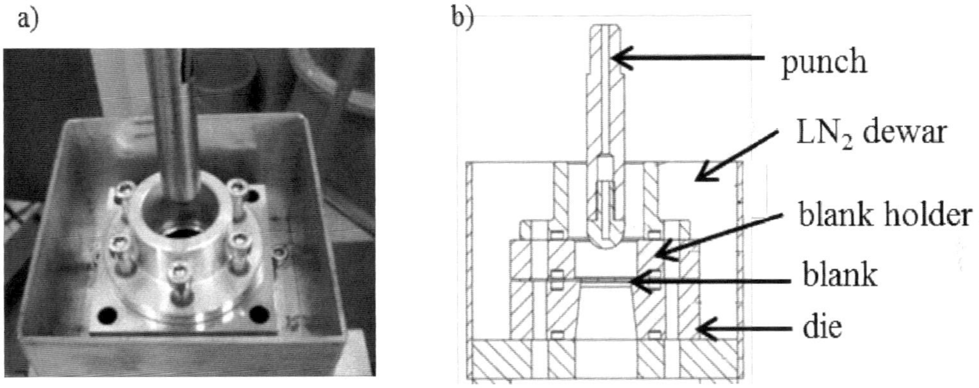

Figure 2. a) Low temperature stretch forming apparatus; b) Cross-sectional view of LDH set-up.

3. Results and Discussion

3.1. Effect of temperature on strength and ductility

Figure 3 shows the true stress-strain relationship of AA5182 and AA6016 at different temperatures. It can be seen that a reduction in temperature has a positive effect on the total strain values. The strain values from AA5182, Figure 4a, rise from initially 0.24 at 25°C to 0.43 at -196°C, which is an overall increase of 75 %. A similar behaviour can be noticed for AA6016 (Fig. 3b), where the initial strain of 0.24 at 25°C could be enhanced by 42% at -196°C and reached a total strain value of 0.34. Furthermore, the true stress rises noticeably when temperature is lowered. The phenomenon of optimizing the stress-strain behaviour at low temperatures is based on the suppression of the thermal nature of cross-slip, which results in a higher strain hardening rate and therefore strain values can be increased [11-13].

3.2. Mathematical formulation of the flow stress

Ludwik, Hollomon, and Backofen identified that the flow stress σ_f can be elevated by raising the strain ε as well as lifting the strain rate $\dot{\varepsilon}$ [14]. The general relationship between the physical values is given by the equation:

$$\sigma_f = \sigma_f(\varepsilon, \dot{\varepsilon}) \qquad (1)$$

A relationship which was empirically developed by Ludwik modified by Hollomon and extended by Backofen [15] has the form:

$$\sigma_f = C_0 \cdot \varepsilon^n \cdot \dot{\varepsilon}^m \quad (2)$$

Where C_0 is a material dependent constant (strength coefficient), n being the strain-hardening coefficient and m represents the strain rate sensitivity parameter. At temperatures where a metallurgical softening takes place to a limited extent ($T \leq 0.4T_m$; T_m = melting point [K]), an increase of the strain rate will not affect the true stress. Therefore, the m value can be assumed to be much smaller than 1. If m is set to be 0 then:

$$\sigma_f = C_0 \times \varepsilon^n \quad (3)$$

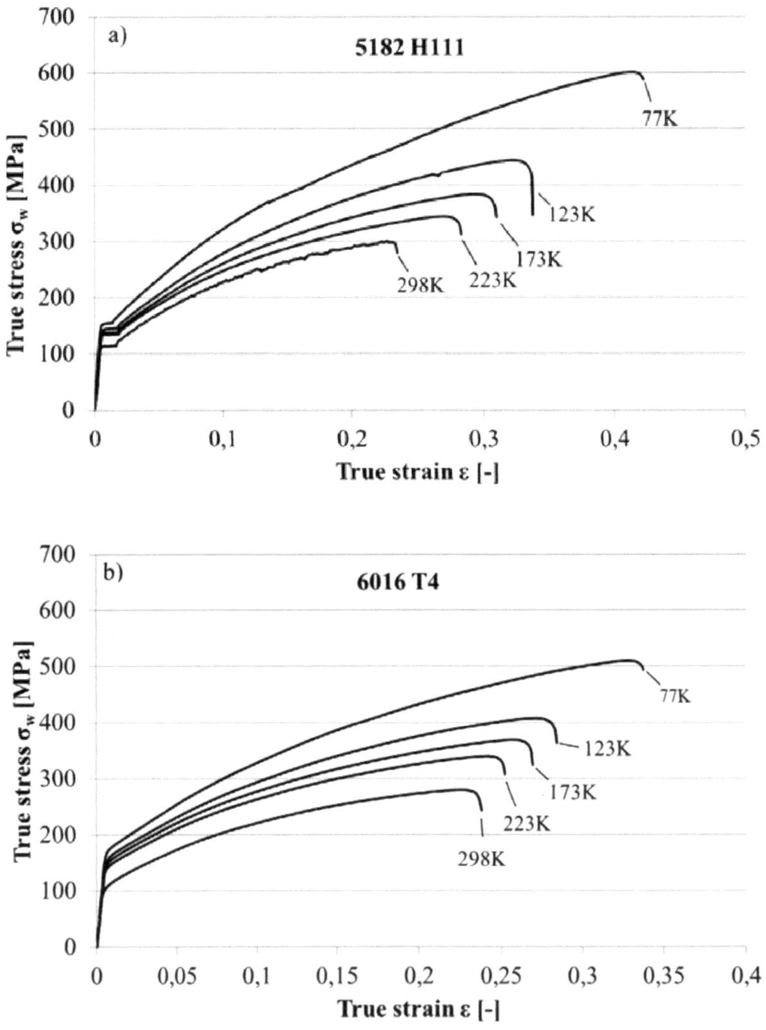

Figure 3. True stress –true strain diagram of a) AA5182 H111 and b) AA6016 T4 at different temperatures.

Using this relationship most stress-strain curves can be described accurately in the region of uniform elongation at $T \leq 0.4T_m$. As exemplified in Heiser [16] the strain hardening coefficient n can be written as true uniform strain ε_u or related to the uniform elongation e_u.

$$n = \varepsilon_u = ln(1 + e_u) \tag{4}$$

The strain hardening coefficient n describes to what extent a material can be deformed before localised necking will start, *i.e.* the flow curve. Figure 4 shows correlations between experimental evaluated strain hardening coefficients using Equation (4) and uniform elongation e_u in relation to temperature of EN AW-5182 H111 and EN AW-6016 T4 specimens. In addition, the Ludwik curve is included which represents n calculated using Equation (4).

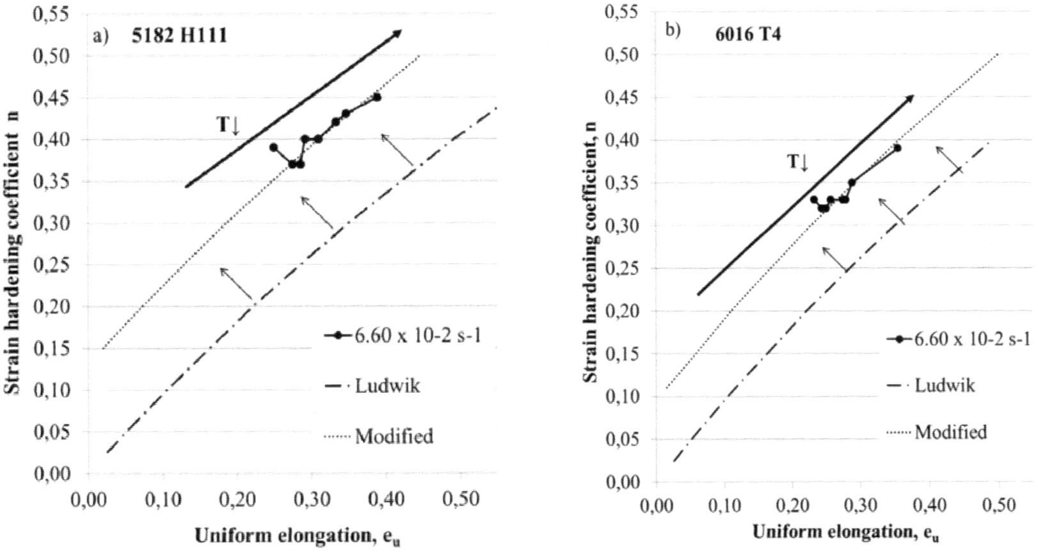

Figure 4. Strain hardening exponent n and uniform elongation e_u for a) AA5182 H111 and b) AA6016 T4. Additionally included is the dependence of strain hardening coefficient and uniform elongation after Ludwik if only strain hardening is considered.

Both aluminium alloys show an increase of the strain hardening coefficient and uniform elongation as temperature is reduced, Figures 4a and 4b. Furthermore, a parallel offset from the initial Ludwik curve can be noticed. Thus, Equation (3) seems not to be precisely enough for tested aluminium alloys. A modification to this equation is indicated by Equation (5), where constant p is added, which provides a shift of the Ludwik curve shown in Figures 4a and 4b.

$$n = ln(1 + e_u) + p \tag{5}$$

If Equation (5) is put into Equation (3) it follows that:

$$\sigma_f = C_0 \times \varepsilon^{n-p} \tag{6}$$

Figure 5 shows the correlation of experimental obtained flow curves (EN AW-5182 and EN AW-6016) and the respective flow curve approximations using the extended formulation, Equation (6).

The constant p is determined by using a target value search algorithm, where the error total of the flow curve approximation is minimised.

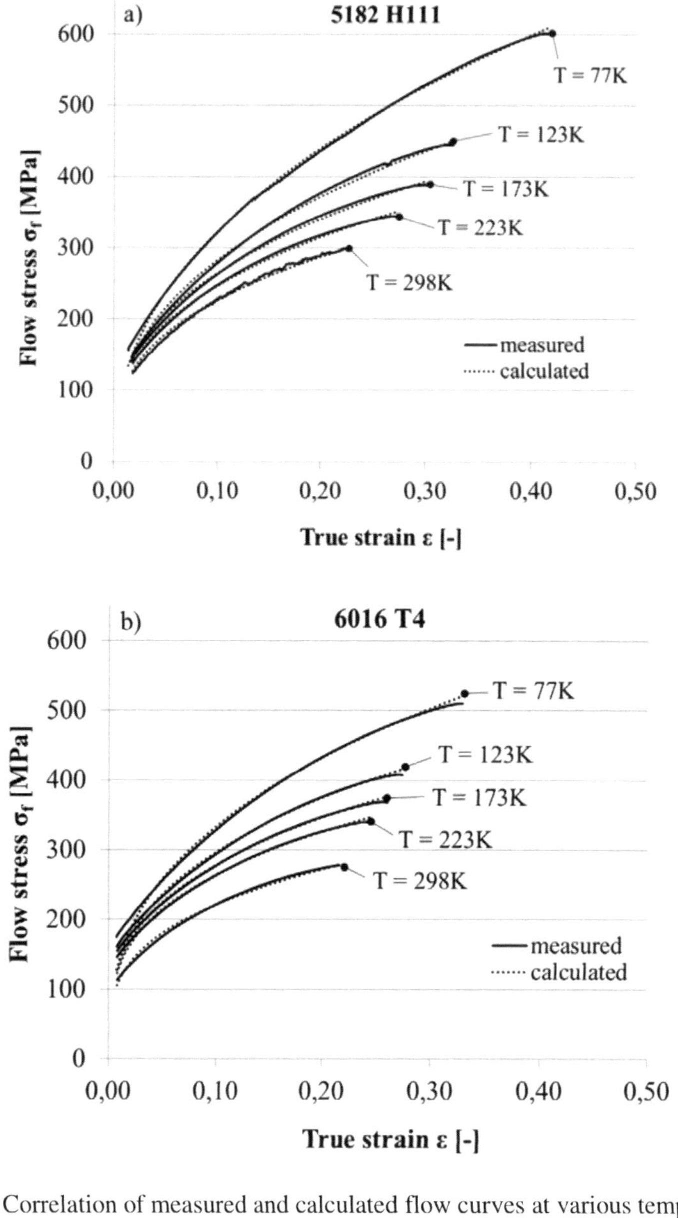

Figure 5. Correlation of measured and calculated flow curves at various temperatures for a) AA5182 H111 and b) AA6016 T4.

3.3. Limiting dome height (LDH) D_h

Figure 6 shows the relationship between limiting dome height and temperature for tested aluminium alloys. Similar to the tensile tests the EN AW-5182 shows an increase in formability if temperature is decreased. From 298 K to 173 K there is no significant change of the dome height. The D_h value rises notably between 173 K and 77 K. The initial dome height at room temperature can be increased by 20% when the EN AW-5182 material is exposed to 77 K. A different behaviour, concerning LDH, shows the EN AW-6016 material when temperature is lowered. It can be seen that the formability behaviour declines between 298 K and 123 K. Only at 77 K the initial D_h value could be surpassed by around 2%, therefore, EN AW-6016 shows limitations in the formability at hydrostatic stress conditions when temperature is decreased.

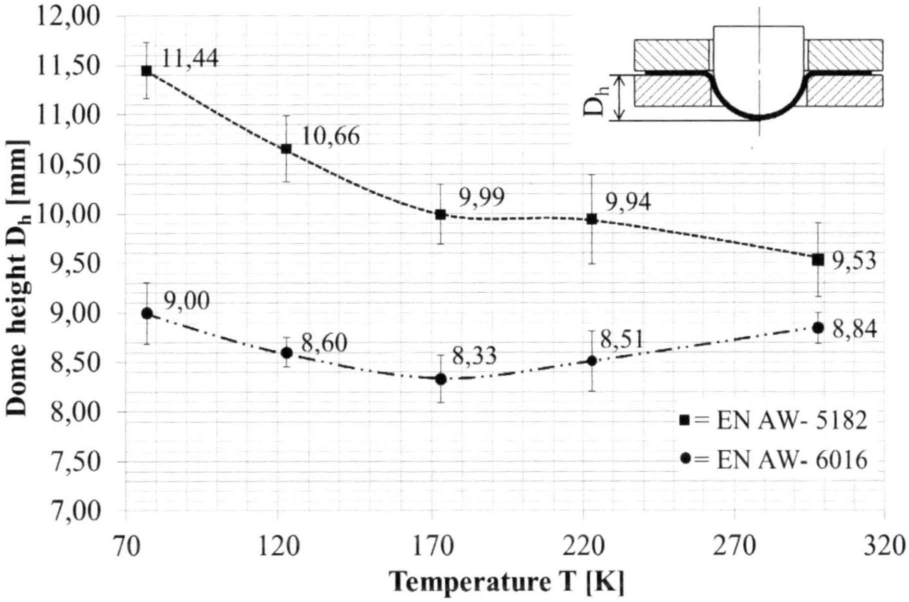

Figure 6. Limiting dome height of EN AW-5182 and EN AW-6016 at different temperatures.

4. Conclusion

Tensile testing at low temperatures showed an increase of elongation values for both tested aluminium alloys EN AW-5182 H111 and EN AW-6016 T4. Corresponding stress values rise as temperature is decreased. The flow curves obtained at sub-zero temperatures can be approximated quite accurately by an adapted stress-strain relationship based on the model proposed by Ludwik and Hollomon. LDH tests showed that the stretch formability of EN AW-5182 can be enhanced at temperatures below 298K which is comparable to the conducted tensile tests. A different behaviour than that found in uniaxial tensile testing can be noticed for EN AW-6016 for biaxial stress conditions as temperature was lowered, which limits the low temperature potential for this alloy.

References

[1] Wang H, Luo Y-B, Friedman P, Chen M-H and Gao L 2012 *Trans. Non. Met. Soc.* **22** 1-7.
[2] Dössel K-F 2013 *J. Oberfl.* **53-3** 20-24.
[3] Pink E and Grinberg A 1984 *Aluminium* **60** 764.
[4] Pink E and Grinberg A 1984 *Aluminium* **60** 687.
[5] Hogg M 2006 *Herstellung und Umformung lokal wärmebeahndelter Platinen* vol 53, ed Siegert K and Liewald M, DGM Informationsgesellschaft mbH, Frankfurt.
[6] Reed R P and Clark A F 1983 *Materials at low temperatures*, ASM, Ohio.

[7] Ekin J W 2007 *Experimental Techniques for Low-Temperature Measurements,* University Press, Oxford.
[8] Aluminium standards and data 1976 Congress no **74**-83459, Aluminium Association, New York.
[9] DIN ISO 6892-1 2009 *Metallische Werkstoffe - Prüfverfahren bei Raumtemperatur*, Beuth-Verlag, Berlin.
[10] DIN ISO 20482 2013 *Metallische Werkstoffe - Bleche und Bänder - Tiefungsversuch nach Erichsen,* Beuth-Verlag, Berlin.
[11] Wigley D A 1971 *Mechanical Properties of Materials at Low Temperatures,* Plenum, New York.
[12] Seeger P H A 1958 *Phil. Mag.* **3** 470.
[13] Nabarro F R N 1989 *Act. Met.* **37** 1521-1546.
[14] Gregory J K 1983 *Superplastic deformation in oxide dispersion strengthened nickel base superalloys,* University Press, Stanford.
[15] Heine B 2011 *Werkstoffprüfung Ermittlung von Werkstoffeigenschaften* (München: Carl Hanser).
[16] Heiser M and Lange G 1992 *Z. Met.***83** 115

IOP Conf. Series: Materials Science and Engineering **74** (2015) 012015 doi:10.1088/1757-899X/74/1/012015

Numerical prediction of the low-velocity impact damage and compression after impact strength of composite laminates

Wei Tan[1], Brian G. Falzon[1], Louis N. S. Chiu[2], Mark Price[1]

[1] School of Mechanical and Aerospace Engineering, Queen's University Belfast, Ashby Building, Belfast, BT9 5AH, UK.

[2] Department of Mechanical and Aerospace Engineering, Monash University, Building 31, Clayton, VIC 3800, Australia.

b.falzon@qub.ac.uk

Abstract. Low-velocity impact damage can drastically reduce the residual mechanical properties of the composite structure even when there is barely visible impact damage. The ability to computationally predict the extent of damage and compression after impact (CAI) strength of a composite structure can potentially lead to the exploration of a larger design space without incurring significant development time and cost penalties. A three-dimensional damage model, to predict both low-velocity impact damage and compression after impact CAI strength of composite laminates, has been developed and implemented as a user material subroutine in the commercial finite element package, ABAQUS/Explicit. The virtual tests were executed in two steps, one to capture the impact damage and the other to predict the CAI strength. The observed intra-laminar damage features, delamination damage area as well as residual strength are discussed. It is shown that the predicted results for impact damage and CAI strength correlated well with experimental testing.

1. Introduction

Composite materials are increasingly used in the aerospace and automotive industries due to their high specific stiffness and strength, nevertheless, the susceptibility to damage from low-velocity impact event (*i.e.* falling tools, runway debris *etc.*) is one of the major design concerns. When subjected to impact loading, the composite structure can be degraded through various failure mechanisms including matrix cracking, fibre breakage and delamination. Impact damage can dramatically reduce the residual mechanical properties of the structure even with barely visible impact damage. It is therefore essential to develop a reliable tool for the prediction of the impact damage process and evaluation of corresponding residual strength.

A limited number of studies on predicting the CAI strength of damaged composite structures are available in the literature. Uda *et al.* [1] investigated failure mechanisms of impact-damaged UT500/Epoxy and AS4/PEEK CFRP laminates subjected to compression fatigue. Ghelli and Minak [2] conducted CAI tests on thin laminates, taking into account sub-laminate buckling under compression. Most of the current work was performed on composites with predefined delaminations or imperfections [3]. A model where both virtual tests (impact and CAI) are performed on composite sandwiches can be found in Davies *et al.* [4]. González *et al.* [5] proposed a 3D FE model with inter-laminar and intra-laminar damage using a rigorous thermodynamic framework for drop-weight impact and CAI test simulations and studied the effect of different stacking sequences. The numerical results showed large oscillations compared to experimental data and took 12-15 days to execute, which is very computational expensive. Rivallant *et al.* [6] presented a numerical model for impact and CAI using the interface element between neighbouring volume elements to model intra-laminar matrix cracking, cutting the run-time to 12-15 hours.

The present work, based on an extension of a previous model by Falzon *et al.* [7-10], was used to simulate both impact and CAI. The results of the simulation are compared with experimental data. The observed intra-laminar damage features, as well as residual strength prediction are discussed. It is shown that the predicted results for impact damage and CAI strength correlated well with experimental testing.

2. Materials and Methods

2.1. Composite Damage Model

The degraded damage model captures failure modes in the forms of matrix cracking, fibre breakage and delamination as shown in Figure 1. Details of the damage model may be found in [7-9] which accounts for these damage modes in a laminate subjected to a 3D stress state.

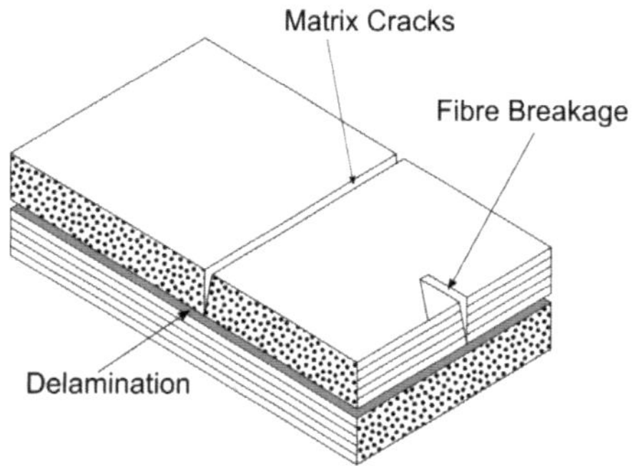

Figure 1. Damage Modes of Composite Laminates.

2.2. CAI Test Model

In this context, the test model is the same as in reference [6]. The virtual tests set up in ABAQUS 6.12/Explicit were executed in two steps, one to capture the impact damage and the other to predict the CAI strength. The laminate plate measured 100 mm × 150 mm × 4.16 mm, simply-supported within a 'picture frame' with a 75 mm × 125 mm effective test section (Figure 2). The T700/M21 composite plate was impacted with 29.5 J of impact energy using a hemi-spherical, 16 mm diameter, 2 kg impactor.

Once the simulation of the impact process was completed, the CAI simulation was subsequently carried out by constraining the out-of-plane direction of the nodes in contact with the picture frame. The planar displacements were also constrained for all nodes at the bottom of the laminate in contact with the support. Although the CAI test is essentially quasi-static (0.5mm/min), it was simulated by using ABAQUS/Explicit to avoid the severe convergence difficulties encountered with implicit analysis. The loading speed was chosen at 3.75 m/min to reduce the CPU running time. Selective mass scaling (scales only elements whose stable time increment is below the value assigned to a time increment of 1e-07) was also employed in the CAI process to achieve a reasonable run time.

2.3. Material Properties

Material properties were obtained from the literature [6], given by Table 1. The $c_i (i = 1,2,3)$ describe a quadratic constitutive shear stiffness curve obtained using a least square fitting method [11]. $\Gamma_{ii}^{Mode} (i = 1,2 \text{ and Mode} = \text{Tensile, Compressive})$ denote the intra-laminar fracture toughness in the longitudinal (11) and transverse (22) directions, and G_I and G_{II} are the inter-laminar fracture toughness for mode I and mode II.

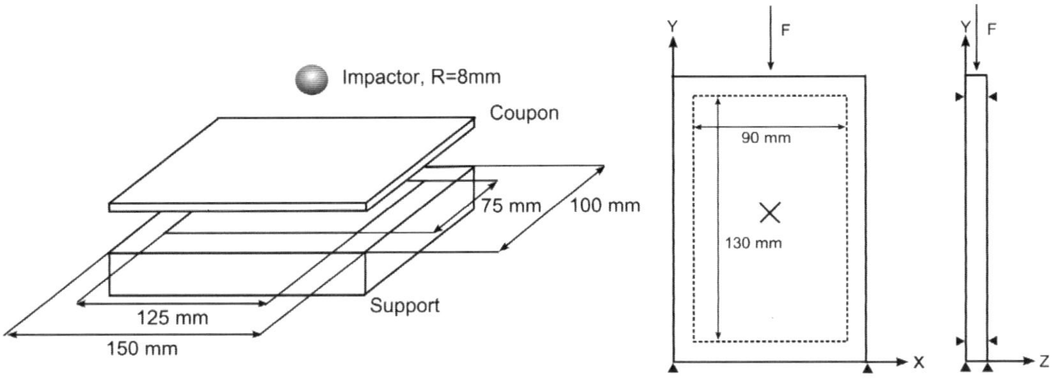

Figure 2. Impact test and CAI setup with the boundary condition.

Table 1. Material Properties of T700/M21 composite.

Property	Values	Property	Values	Property	Values
E_1	130 GPa	X^C	1250 MPpa	Γ_{22}^C	1.6 N/mm
$E_2 = E_3$	7.7 GPa	Y^T	60 MPa	$\Gamma_{12,C} = \Gamma_{23,C}$	1.6 N/mm
$\nu_{12} = \nu_{13}$	0.3	Y^C	290 MPa	G_I	0.5 N/mm
ν_{23}	0.35	S_{12}	110 MPa	G_{II}	1.6 N/mm
$G_{12} = G_{13}$	4.8 GPa	Γ_{11}^T	133 N/mm	c_1	37833 MPa
G_{23}	3.8 GPa	Γ_{11}^C	10 N/mm	c_2	16512 MPa
X^T	2080 MPa	Γ_{22}^T	0.5 N/mm	c_3	2334.3 MPa

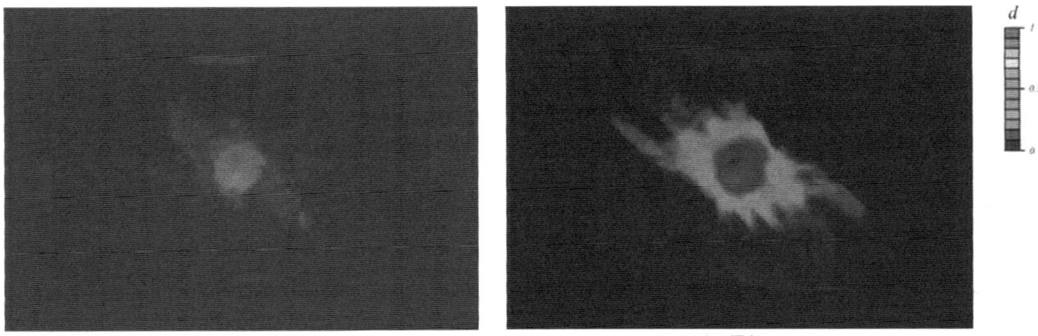

(A) (B)

Figure 3. (A) Matrix Damage, and (B) Delamination of 29.5 J

3. Results

3.1. Impact Matrix Damage

The intra-laminar damage model allowed the various forms of intra-laminar failure to be investigated as the impact event progressed. Figure 3 shows the superposition of the matrix cracking throughout the layers. Matrix cracking was found to be the dominant form of intra-laminar failure. Matrix damage mainly occurred in the centre part of the impact area with a symmetric distribution similar to that reported in [12]. It is clearly that the matrix damage propagated from the centre and towards the boundary of "picture frame". The peanut shaped damage contour for the delamination can be observed in Figure 3 which matches well with the experimental results from C-scans [6], though the size of the damage area is smaller. These could be attributed to the fact that the experimental results are the combination of the matrix damage and delamination.

3.2. Force-Time and Force-Displacement Curves

In Figure 4A, the experimental results of force-time history are not available and the numerical results show the overall impact response. The load oscillation is an artefact of the numerical method due to matrix failure and delamination failure. Upon contact with the laminate, the impactor force increased gradually with some fluctuation before it reached the peak load when impact velocity was zero. From this stage onwards, the force began to drop until it reached zero, during which the impactor rebounded. The force-displacement history of the impactor correlated well with the experimental results shown in Figure 4B. The initial contact response and maximum displacement of the impactor are captured accurately compared to experimental results [6].

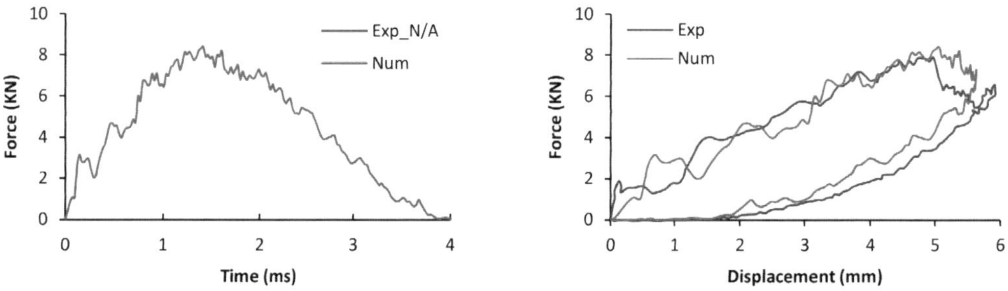

(A) (B)

Figure 4. (A) Force-Time History, and (B) Force-Displacement History of 29.5 J

3.3. CAI Matrix Damage

The extensive matrix damage of the CAI test under impact energy of 29.5 J is shown in Figure 5. Most matrix damage occurred at the centre line through the indented zone. The CAI test process from (A) to (D) indicated that new matrix damage initiated from the two outer edges, possibly due to free edge effects, and propagated towards the damaged centre of the panel. Sub-laminate buckling is also observed in (E).

3.4. CAI Stress-Displacement Curve

CAI experimental results in terms of the stress-displacement relationship were only available for the 29.5 J impact case [6]. Applied stress versus end displacement curve in Figure 6 shows that good correlation was achieved between the experimental and numerical results, with the ultimate load being predicted to within 9.3% of the experimental results.

Another model (undamaged line) was executed with no impact damage which shows that the damaged panel failed at 40% of the load carried by the pristine panel.

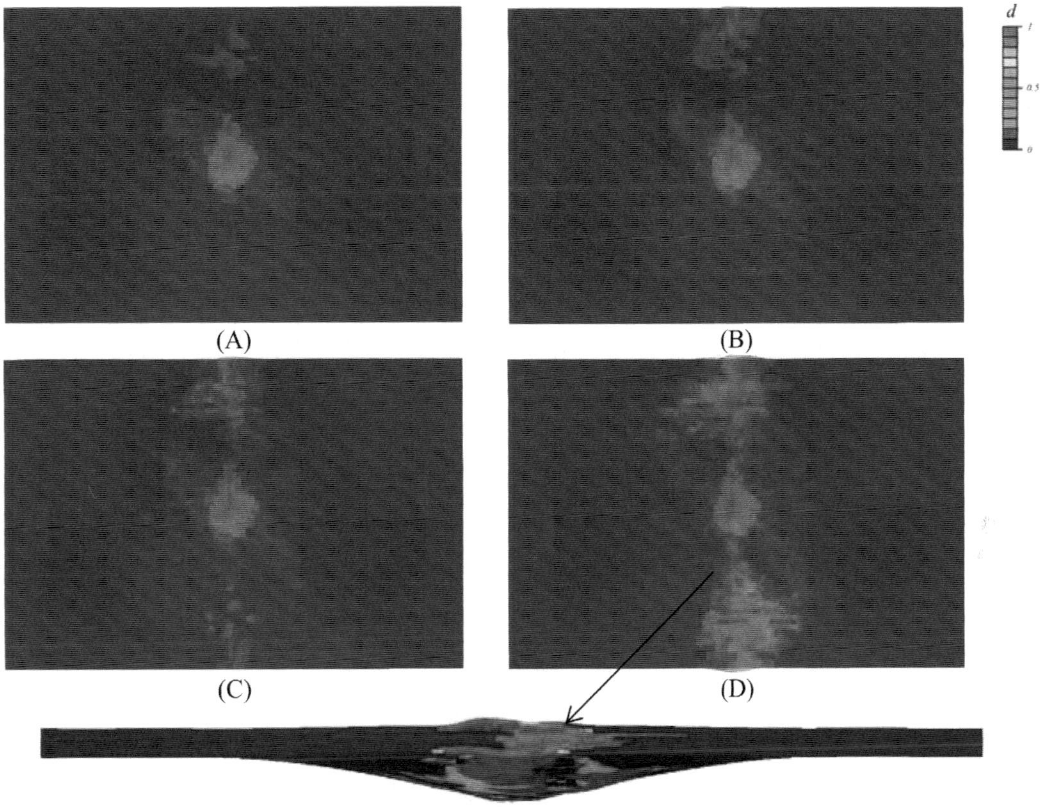

Figure 5. CAI Matrix Damage, (A) 14 ms, (B) 15 ms, (C) 16 ms, (D) 22 ms, (E) Side view.

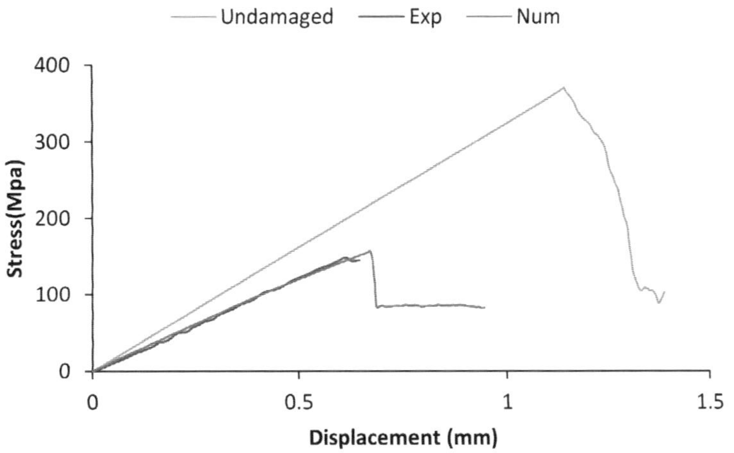

Figure 6. CAI stress-displacement curve.

4. Conclusions

Through the data obtained from the force-displacement and damage parameters, the impact test successfully demonstrated that the model can capture intra-laminar and inter-laminar damage well – both qualitatively and quantitatively – for the given impact energy. Concerning the CAI simulation, the observed intra-laminar damage features, as well as residual strength were accurately predicted. This was achieved without the need to calibrate the input parameters. The current model has proved to be a reliable and efficient tool to numerically predict the CAI strength.

Future work will focus on extending this computational damage model to capture high energy crush events. This will enable the virtual testing of bird strikes and assessment of crashworthiness of aero-structures.

Acknowledgements

The authors would like to gratefully acknowledge the funding from Queen's University Belfast/China Scholarship Council (QUB/CSC) PhD Scholarship and the support of High Performance Computing (HPC) systems of QUB. The corresponding author would like to acknowledge the financial support of Bombardier and the Royal Academy of Engineering.

References

[1] Uda N, Ono K, Kunoo K, 2009, "Compression fatigue failure of CFRP laminates with impact damage." *Composites Science and Technology* **69** 2308-14.

[2] Ghelli D, Minak G, 2011, "Low velocity impact and compression after impact tests on thin carbon/epoxy laminates." *Composites Part B: Engineering* **42** 2067-79.

[3] Yan H, Oskay C, Krishnan A, Xu LR, 2010, "Compression-after-impact response of woven fiber-reinforced composites." *Composites Science and Technology* **70** 2128-36.

[4] Davies GAO, Hitchings D, Besant T, Clarke A, Morgan C, 2004, "Compression after impact strength of composite sandwich panels." *Composite Structures* **63** 1-9.

[5] González E, Maimí P, Camanho P, Turon A, Mayugo J, 2012, "Simulation of drop-weight impact and compression after impact tests on composite laminates." *Composite Structures* **94** 3364-78.

[6] Rivallant S, Bouvet C, Hongkarnjanakul N, 2013, "Failure analysis of CFRP laminates subjected to compression after impact: FE simulation using discrete interface elements." *Composites Part A: Applied Science and Manufacturing* **55** 83-93.

[7] Falzon BG, Apruzzese P, 2011, "Numerical analysis of intralaminar failure mechanisms in composite structures. Part I." *FE implementation Composite Structures* **93** 1039-46.

[8] Falzon BG, Apruzzese P, 2011, "Numerical analysis of intralaminar failure mechanisms in composite structures. Part II." *Applications Composite Structures* **93** 1047-53.

[9] Faggiani A, Falzon BG, 2010, "Predicting low-velocity impact damage on a stiffened composite panel." *Composites Part A: Applied Science and Manufacturing* **41** 737-49.

[10] Chiu LN, Falzon BG, Boman R, 2013, "A Continuum Damage Mechanics Model for the Analysis of the Crashworthiness of Composite Structures: A work in progress." *Proc. 15th Australian Aeronautical Conference,* Melbourne.

[11] Preetamkumar M, Gilles L, Pierre L, Ana-cristina G, 2009, *50th AIAA / ASME / ASCE / AHS / ASC Structures, Structural Dynamics, and Materials Conference*: American Institute of Aeronautics and Astronautics.

[12] Raimondo L, Iannucci L, Robinson P, Curtis P, 2012, "A progressive failure model for mesh-size-independent FE analysis of composite laminates subject to low-velocity impact damage." *Composites Science and Technology* **72** 624-32.

IOP Conf. Series: Materials Science and Engineering **74** (2015) 012016 doi:10.1088/1757-899X/74/1/012016

Development of a Computer simulation approach for honeycomb constructions for aerospace application

OV Tatarnikov[1,2], KS Karpenkov[2]

[1]Mathematics Department, Plekhanov University of Economics, Moscow, Russia
[2]Bauman University, Moscow, Russia

tatarnikov.ov@rea.ru

Abstract. An approach to definition of a homogeneous simulation model for honeycomb structures has been developed and verified for specimens containing a finite number of cells. The elastic characteristics of the model were evaluated basing on the results of tensile and shear numerical tests of honeycomb specimen. This is an extension of earlier work related with spatially reinforced composites. The simulation model was validated for specimens comprised of different numbers of cells in the specimen to expose the scale effect influence. As the number of cells was increased, the calculated values of the moduli E_x and E_y converged, confirming the theoretical result that the appropriate model is transversely isotropic rather than orthotropic for the honeycomb specimen investigated. Elastic properties obtained from the numerical test of the honeycomb structure were then applied in the characterization of continuous medium. The examination was carried out using criteria expressing basic features of homogeneous body. The case of a honeycomb integrated with composite plates as a sandwich structure was analysed for a complex loading case as well as buckling and eigen-frequency analysis.

1. Introduction

There are two prevailing approaches to the analysis of constructions containing honeycomb panels. According to one the real geometrical configuration of a honeycomb is considered, *i.e.* a honeycomb structure is presented as a set of hexagonal cells [1-3]. That type of a model can be applied to the simulation of a construction of a simple configuration. In dealing with aerospace parts, it is often necessary to take into account an ensemble of several elements made of different types of materials. In this case the consideration of each individual cell in the analysis of a complex construction containing honeycomb panels becomes difficult and impractical.

In such cases, a honeycomb structure can be simulated by an equivalent continuous medium, with adequate elastic characteristics. In accordance with such an approach to the simulation of a honeycomb structure, formulas based on the standard beam theory are usually used for averaging the elastic characteristic [4]. It should be noted that the beam approach gives the transversal isotropic material for modeling of a honeycomb. Such an approach is based on boundary symmetry conditions and therefore assumes an infinite number of regular cells: it does not duly take into account certain geometrical and mechanical features of a honeycomb construction, for instance irregularities in the hexagonal geometry, tolerance variations in shape and wall thickness, and the presence of construction features such as weld lines or material crimping, nor does it take in account with the finite size ratio of a honeycomb construction and a cell, or anisotropy of mechanical properties of a honeycomb core material. All of this has led to the development of the numerical model that would adequately simulate the honeycomb structure. The proposed approach is an extension of a previously work related with the modelling of spatially reinforced composites [5].

2. Homogeneous model of honeycomb

In problems of mechanics concerning materials containing regularly distributed heterogeneities, for instance, composites, porous materials or honeycombs the homogeneous continuous model is to be chosen. Mechanical properties of such a model are determined either by means of mechanical tests or using appropriate averaging methods. Such a model has to meet three criteria:

1. A model and a prototype must have the same type of symmetry concerning mechanical properties.
2. Potential energy of the model and prototype must be equal under the same loading condition.
3. Mass of the representative volume of the model and prototype must be equal.

In a regular hexagonal cell honeycomb structure elastic symmetry axes of order 6 are found passing through the finite number of points, exactly through the center of each cell (Figure 1).

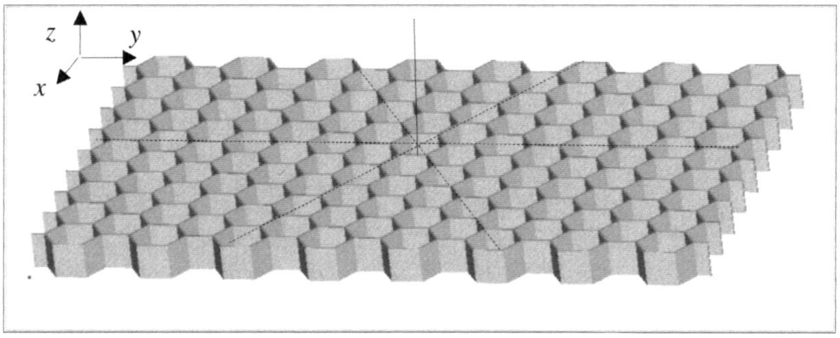

Figure 1. The elastic symmetry axis of order 6 passing through the honeycomb

It is known that a material through each point of which there passes an axis of elastic symmetry of the 6th order is a transversely isotropic one [6]. The number of such axes in a honeycomb is limited by the number of cells in the structure. That means that a transversal isotropic material can be considered as a honeycomb homogeneous model in the cases when the characteristic dimension of a cell much less compeering with the characteristic dimension of a construction. For transversely isotropic materials there are 5 independent elastic constants. Where there are only a small number of honeycomb cells, where the shape is not a regular hexagon, or where geometry variation has to be taken into account, the symmetry required for transverse isotropy breaks down. In this case the elastic properties of a honeycomb model are fully determined by 9 independent elastic constants,

- Elastic moduli – $E_x, E_y, E_z,$
- Shear moduli – $G_{xy}, G_{xz}, G_{yz},$
- Poisson's ratios – $\nu_{xy}, \nu_{xz}, \nu_{yz}.$

but it might be anticipated that for honeycombs displaying only minor symmetry violations, some of independent constants would have very similar values.

The second criterion for a homogeneous model can be presented by a relation that establishes the equality of potential elastic energy of a honeycomb and that of a model

$$\iint_{S_h} p_i u_i ds = \iiint_{V_m} \sigma_{ij}\varepsilon_{ij}\, dv. \tag{1}$$

In (1), the left side expresses the work of the surface load p_i applying on the boundary surface S_h of a honeycomb when the displacement is u_i. The right side presents the elastic energy of the volume V_m of the model.

For the determination of elastic characteristics of the honeycomb two types of numerical tests based on solid FE models were carried out. Elastic moduli and Poisson's ratios were calculated using data of honeycomb tension test simulation (Figure 2). For instance, the E_x modulus was evaluated as

$$E_x = \frac{\sigma_x}{\varepsilon_x} = \frac{F_x}{S_x tg\theta}\ , \tag{2}$$

where $\sigma_x = F_x/S_x$ is the normal stress, F_x the resultant tensile force, S_x the effective area of honeycomb end face, strain $\varepsilon_x = \partial u_x/\partial x = tg\theta$ is the strain, and θ the angle between straight line representing the relationship between displacement u_x and the x coordinate (Figure 3).

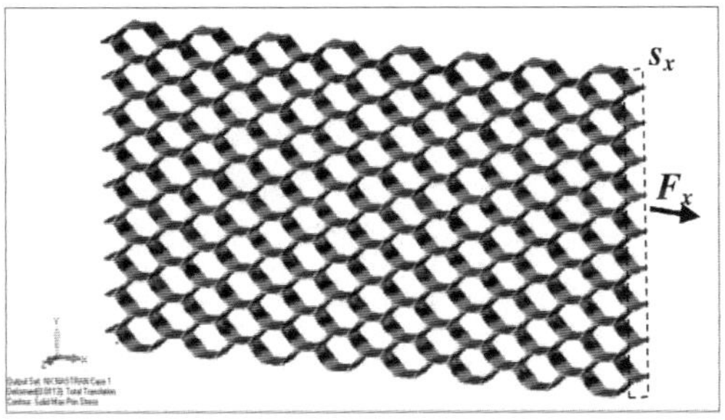

Figure 2. Finite element model (solid elements) of honeycomb under tension. Deformed style.

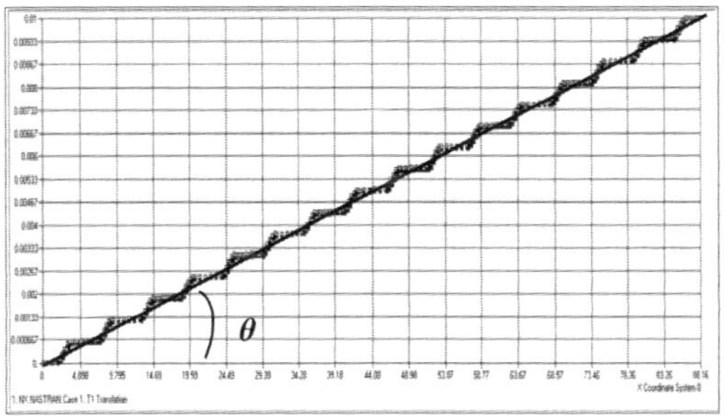

Figure 3. Displacement u_x vs x coordinate.

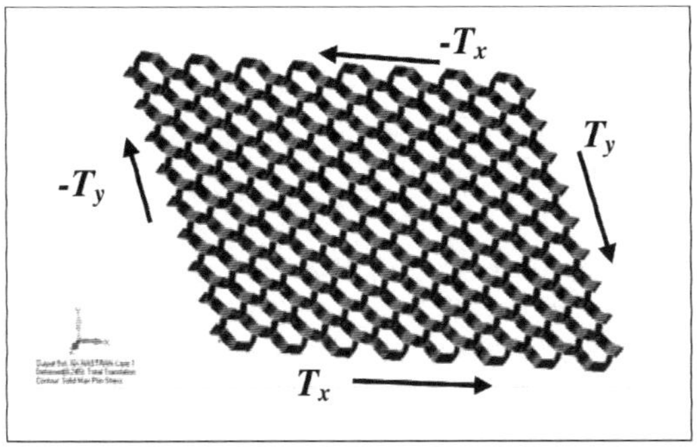

Figure 4. Finite element (solid elements) evaluation of shear modulus G_{xy}. Deformed style.

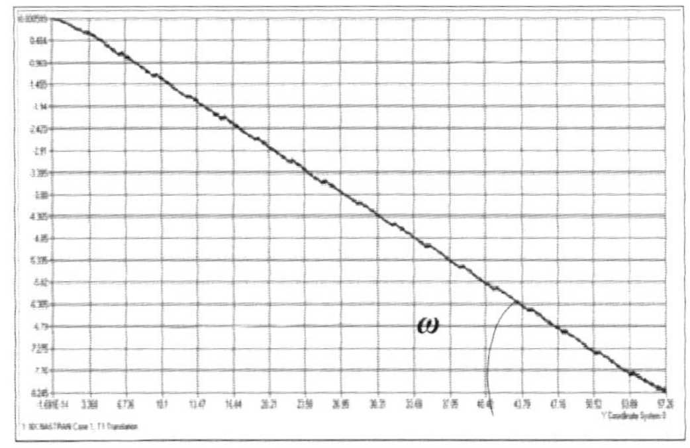

Figure 5. Displacement u_x vs y coordinate.

Shear moduli were evaluated basing on the data obtained from the simulation of pure shear test (Figure 4). For example, shear modulus G_{xy} was calculated as

$$G_{xy} = \frac{\tau_{xy}}{\gamma_{xy}} = \frac{T_y}{S_x tg\omega} \tag{3}$$

where $\tau_{xy} = T_y/S_x$ is the shear stress, T_y the resultant tangential force applied on the S_x face of the honeycomb specimen, $\gamma_{xy} = \partial u_x/\partial y = tg\omega$ the shear strain and ω the angle between the straight line representing the relationship between displacement u_y and y coordinate (Figure 4).

It should be noted that relations (2) and (3) follow from (1) under the assumption that the loads are uniformly distributed over the surfaces of a honeycomb.

3. Honeycomb model

As an example, an aluminum honeycomb was considered. Table 1 contains the specifications defining the mechanical properties of the honeycomb core material and cell dimensions.

Table 1. Honeycomb specifications

Young's modulus (MPa)	70000
Poisson's ratio	0.3
Cell wall thickness (mm)	0.3
Cell wall height (mm)	5
Cell wall length (mm)	3.5

The development of a computational model included the choice of the number of cells in the specimen. The graph shown in Figure 6 exhibits the scale effect in the determining of honeycomb elastic moduli.

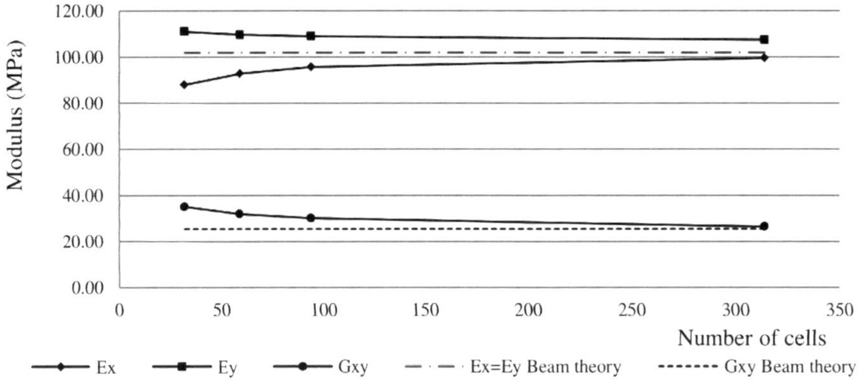

Figure 6. Scale effect in the determining of elastic moduli E_x, E_y and G_{xy}

The observed variation of the difference between moduli E_x and E_y shows the tendency of conversion from orthotropic to transversely isotropic model as number of cells increases. The influence of scale factor that can be determined as ratio of characteristic specimen size L to characteristic cell size l is illustrated by data presented in the Table 2.

Table 2. Elastic moduli vs scale factor *L/l*

N	L/l	E_x	E_y	E_x=E_y Beam theory	G_{xy}	G_{xy} Beam theory
32	9.5	88.0	111.0	101.8	35.1	25.5
59	13.0	92.8	109.6	101.8	31.9	25.5
94	16.4	95.7	108.9	101.8	30.1	25.5
314	30.4	99.5	107.4	101.8	26.5	25.5

Elastic properties of homogeneous model which have been obtained using a set of numerical tension and shear tests of the honeycomb specimens are presented in Table 3.

Table 3. Elastic properties of aluminum honeycomb

Elastic moduli (MPa)						Poisson coefficients		
E_x	E_y	E_z	G_{xy}	G_{xz}	G_{yz}	ν_{xy}	ν_{zx}	ν_{zy}
99.5	107.4	6446.4	26.5	55.9	50.1	0.92	0.005	0.300

It should be noted that all of these data were obtained from tests that were carried out with a heterogeneous honeycomb structure, however, they were used as characteristics of a continuous medium, that is, a homogeneous model of honeycomb. Therefore, it was necessary to verify their suitability for the characterization of the continuous medium. The verification of the homogeneous model developed was performed using criteria expressing well-known features of homogeneous orthotropic and transversely isotropic media [6]. The verification data are presented in Table 4. This analysis thereby establishes the level of orthotropy that arises as a result of the finite size of the specimen, so that where more realistic honeycomb geometries are analysed using the same methodology, it will be possible to differentiate between scale effects and asymmetry effects.

Table 4. Verification of model

Quantity	Value	Criterion
ν_{yx}/E_y	0.00929	$\frac{\nu_{yx}}{E_y}=\frac{\nu_{xy}}{E_x}$
ν_{xy}/E_x	0.00923	
$\nu_{yx}+\nu_{yz}+\nu_{zx}$	1.272	<1.5
$1-\nu_{xy}\nu_{yx}-\nu_{yz}\nu_{zy}-\nu_{xz}\nu_{xz}-2\nu_{yx}\nu_{zy}\nu_{xz}$	0.031	>0

The results confirm the adequacy of the homogeneous model of a honeycomb in the case of simple loading. It was of interest to verify the adequacy of this model for a honeycomb integrated with other construction elements as well as for the case of complex loading, in problems of buckling and dynamics.

4. Honeycomb sandwich structures

Honeycomb sandwich structure is a composition of two skins, made of laminate composites, separated by a honeycomb. These structures are widely used in aerospace items such as wings and fuselage parts, interior elements, etc. due to their high specific stiffness. There are three main types of problem in the design of honeycomb sandwich structures for aerospace application:

1. The bending stresses analysis of the sandwich structure.
2. The buckling analysis of honeycomb sandwich structure.
3. The eigenfrequencies spectrum analysis of the sandwich structure.

In all three cases, it must be understood that where the overall sandwich structure is being considered, and therefore where a homogenized material modelling approach is adopted, the detail of the deformation of the honeycomb and skin is not apparent. In this work both analysis approaches were adopted, and the results compared.

4.1. Bending

An idealized sandwich panel consisting of two carbon skins and aluminum honeycomb, subjected to bending loading was considered. The honeycomb properties were assumed to be the same as in the Table 1. The specifications of carbon skins are presented in Table 5. The scale factor of the sandwich specimen was taken as 10.9.

Table 5. Carbon skins specifications

E_x (GPa)	150
E_y (GPa)	10.7
G_{xy} (GPa)	5.7
ν_{xy}	0.24
Skin thickness (mm)	1.0
Skin length (mm)	88
Skin width (mm)	38

The FE model of the sandwich panel contained solid elements for modelling of the honeycomb and laminate plate elements for modelling of the skins. The ends of the skins were constrained. The bending load was applied as shown in Figure 7. Taking into account the conditions of symmetry, the numerical model was chosen as ¼ part of the structure (Figure 8). Analysis was carried out for two models of the sandwich. In the first one, the honeycomb structure was considered. In the second, the honeycomb was substituted for by a homogeneous model. The skins were modelled explicitly in both cases.

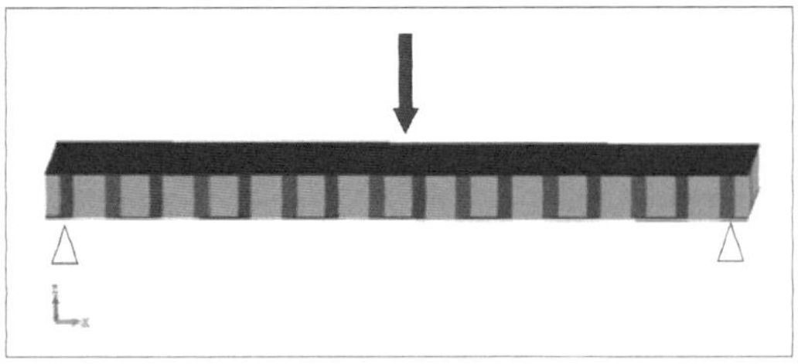

Figure 7. Loading of a honeycomb sandwich.

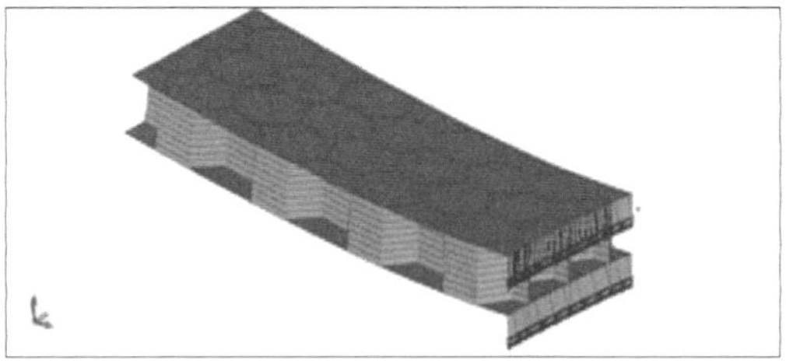

Figure 8. Bending of a honeycomb sandwich plate (¼ part of the heterogeneous model).

IOP Conf. Series: Materials Science and Engineering **74** (2015) 012016 doi:10.1088/1757-899X/74/1/012016

The results of the calculation are presented in Figure 9. There are two curves in the graph that show the distribution of the panel deflections for the two models. Both sets of results coincided to within a tolerance of 2.6%.

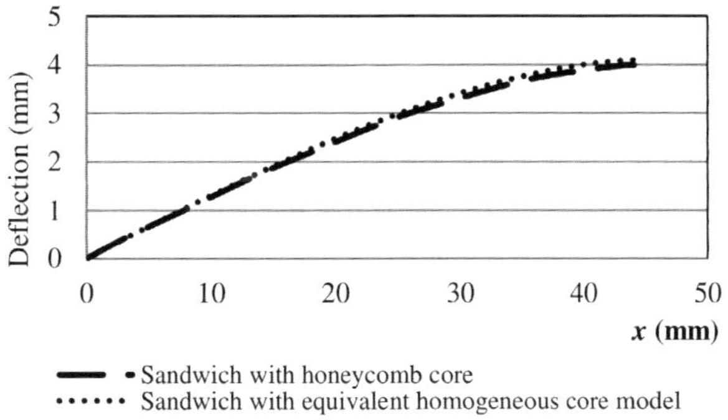

Figure 9. Distribution of deflection along sandwich plate

4.2. Buckling

The same sandwich panel was considered in the buckling problem. On one end of the sandwich both skin plates was loaded by compressive force distributed along skin edges as shown in Figure 10. The other end was constrained. In Table 5, the coefficients of the critical load for both simulation models are presented. Figures 10 and 11 illustrate the 4th buckling mode corresponding to the honeycomb and homogeneous model.

It should be noted that the results in the coefficients and modes match to a sufficient degree.

Table 6. Critical compressive force coefficient

Honeycomb sandwich	Sandwich with homogeneous core
9.716	9.469
39.107	39.029
51.864	52.887
60.958	61.337
70.346	68.213

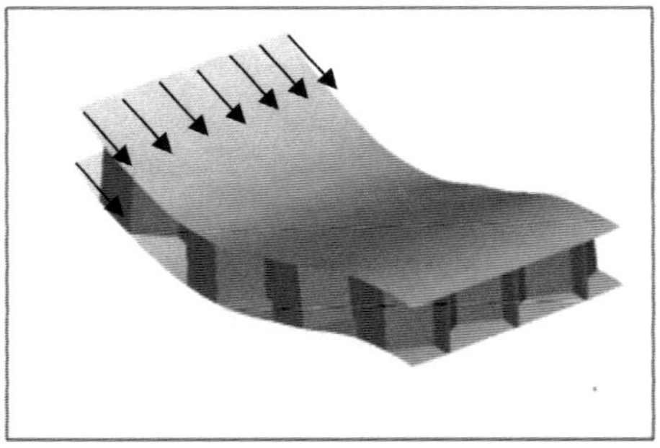

Figure 10. Honeycomb sandwich buckling

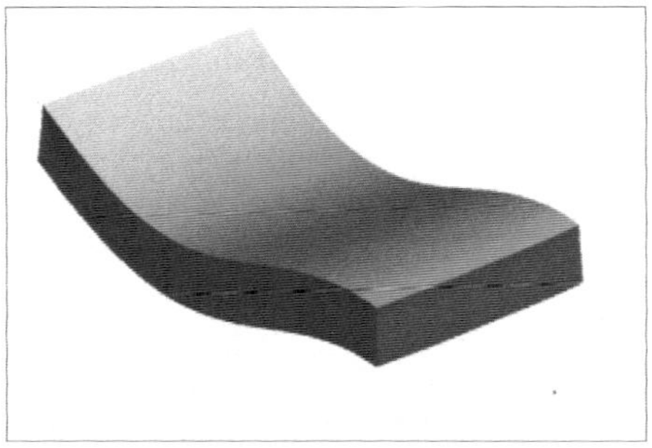

Figure 11. Sandwich model with homogeneous core buckling

4.3. Eigen-frequency
In the eigen-frequency analysis, a sufficient match of the result was reached as in previous cases. In Table 7 the eigen-frequencies corresponding to the first 5 modes are presented.

Table 7. Eigen-frequency (Hz)

Mode	Honeycomb sandwich	Sandwich with homogenous core
1	2.174	2.156
2	13.210	13.104
3	19.772	19.906
4	25.617	25.549
5	26.040	25.971

5. Conclusions
In the case of real engineering materials, there would be significant departure from idealized material properties. The proposed approach allows creating simulation models for static and dynamic analysis of constructions containing honeycomb structures, taking into account the main features of the real materials. It can be applied for honeycomb structures and cellular materials with different type of a core material and cells with a regular and an irregular shape.

References
[1] Masters IG, Evans KE, 1996, "Models for the elastic deformation of honeycombs." *Composite Structures* **35** 403-422.
[2] Florens C, Balmes E, Clero F, Corus M, 2006, "Accounting for glue and temperature effects in Nomex based honeycomb models" *Int. Conf. Noise and Vibration Engineering*, 18-20 September, Louvain, Belgium.
[3] Hu NG, Wang W, Leng GJ, 2012, "Study on Equivalent Modeling of Honeycomb Sandwich Panels for Large Radio Telescopes." *Lecture Notes in Information Technology* **15** *Materials, Mechatronics and Automation*, Nanchang, China, 509-514.
[4] Gibson LJ, Ashby MF, Schajer GS, 1982, "The mechanics of two-dimension cellular materials." *Proc. R. Soc.* **A382** 25-42.
[5] Tatarnikov OV, Gaigarova M, 2012, "Three-Level Design of Composite Structures." *Proc. 2nd Int. Conf. Advanced Composite Materials and Technologies for Aerospace Applications,* 11-13 June, Wrexham, North Wales, UK.
[6] Lekhnitskii SG, 1981, *Theory of elasticity of an anisotropic body,* Mir Publishers.

IOP Conf. Series: Materials Science and Engineering **74** (2015) 012017 doi:10.1088/1757-899X/74/1/012017

Evaluation of resistance of diamond-like carbon coating to the corpuscular radiation in outer space conditions

Elizaveta Tomilova, Valeriy Bashkov, Pavel Mikhalev, Alexander Fedorchenko and Yana Volkova

Bauman Moscow State Technical University, 2-ya Baumanskaya str. 5, Moscow, 105005, Russia

tomilich13@gmail.com

Abstract. The purpose of this work was to research the resistance of thin coatings to the effects of corpuscular radiation, as well as evaluation speed etching of diamond-like films with different content of diamond phase. There were two samples of monocrystalline silicon with DLC coating. To evaluate the resistance, two groups of grooves were etched on each sample. The depth was then measured to calculate a relative etching ratio of DLC coating. The resistance was determined to be four times that of silicon.

1. Introduction

The use of polymers in the structure and devices of spacecraft allows their weight to be reduced, along with their dimensions, thereby increasing their economic efficiency, however, the far lower stability of polymers against open space factors such as vacuum, temperature, and ionizing and particle radiations significantly hinders their wide application. One way to solve this problem is to cover the surfaces of polymers with protective gas tight coatings based on metals and metal oxides [1, 2].

A relatively new trend that is being intensely developed in the recent years is the use of diamond-like carbon (DLC) for the protection of products and materials from various negative factors due to the number of unique properties inherent to those coatings: high mechanical strength, high gas tightness, biocompatibility etc. [3-5]. Over the recent 2-3 years there have been lots of articles in which polymers with DLC layers are considered as promising packaging materials for the food and pharmaceutical industries due to their good gas tightness. This allows one to consider diamond-like coatings as promising for the protection of polymer structures and products used in the open space.

Bashkov *et al* [6] report on the protective properties of DLC coatings under the conditions of open space, demonstrating their gas tightness under thermal and vacuum impacts, and a tangible reduction in material erosion under the impact of particle radiation. This work is a continuation of the study dealing with the compatibility of DLC coatings in the open space, in particular, study of the stability of these coatings against particle flows that cause etching and erosion of most polymers.

The aim of this work was to develop a method for assessing the stability of thin coatings against ion radiations and the etching rate of diamond-like films with various contents of the diamond phase.

2. Test Materials and Experimental Equipment

Silicon was chosen as the material of the substrates for the application of DLC coatings. This choice was accounted for by the fact that during the impact of ion fluxes on samples the etching of the surface layer of different polymers may occur by various mechanisms and at different rates, whereas the etching of single crystal silicon has been well studied and is stable. Furthermore, single crystal silicon is delivered in the form of polished wafers with very high surface purity, while obtaining polymers with comparable surface purity is a complex or potentially impossible task.

In this work samples of DLC coatings were studied that were applied onto 20x20 mm single crystal silicon wafers using the vacuum electric arc method with plasma flow separation and laser arc initiation on an East-01 instrument (OOO NPT, Russia). This coating application method allows synthesizing coatings with high substrate adhesion and provides for a high diamond likeness [7, 8]. The coatings thickness was 170 nm including an adhesive Ti sublayer having a thickness of up to 50 nm.

Before coating, masks were positioned on each sample for forming a window in the coating required for measuring its thickness under a probe microscope. Also this mask was used to form an uncoated area for assessing the impact of the ion flux onto the substrate.

Focused gallium ion beam was used as a source of particle radiation. This radiation was chosen because of the possibility of sharp focusing of the beam and forming a controlled impact area with well-controlled parameters, furthermore, gallium ions are preferable for producing particle radiation because their impact causes physical sputtering of the target material unlike oxygen, nitrogen or fluorine ions that may enter into chemical reactions.

Ion irradiation was provided using the FIB module of the NanoFab-100 ultrahigh vacuum four-chamber nano-technological device (manufactured by CJSC "NT-MTD", Zelenograd) intended for studying the surface of up to 100 mm diameter wafers under the conditions of ultrahigh vacuum and equipped with a gallium ion radiation source.

2.1. Determination of Focused Ion Beam Impact Parameters

Ion beam impact is achieved by multiple scanning of the selected area with a sharp focused ion beam. The main variable parameters of radiation are acceleration voltage, beam width and number of passes. The acceleration voltage and width control the shape (diameter) of the beam and the energy and density of the ion flux, whereas the number of passes determines the time of impact.

For a known etching ratio, these functions allow the time of exposure to be calculated, which is required for etching to the desired depth. Taking into account the high purity of the materials used and their stable parameters, the etching ratios for a wide range of materials used in microelectronics are well known. For single crystal silicon substrates the sputtering ratio for irradiation with 30 keV gallium ions is well known and is 2.412 [9, 10]. The magnitude of ion current depends on the width and acceleration voltage; it cannot be strictly controlled.

3. Experiment

Each sample was exposed to radiation in multiple areas for which the time of impact varied and hence the depths of the etch grooves also differed. Different grooves depth allowed a more accurate determination of etching rate (stability against impact) and assessing the scatter of etching rates depending on depth. The etched areas were at the boundary between the coating and the pure silicon such that the coating boundary was close to the middle of the etched area. This arrangement provided for:

- Simultaneous and similar impact both on the coating and on the silicon
- Good visibility of etched areas during further probe microscope scanning
- Visible comparison between the etching rates of the silicon and the diamond-like coating

Experimental parameters were optimized during method testing. One of the tasks of the work was to determine the size of the etched area. The following criteria were used for selecting these parameters:

- Specific features of surface topology scanning using the probe method
- Minimizing the effect of back-deposition depending on the small size of the area
- Minimizing the time of exposure to the focused ion beam

IOP Conf. Series: Materials Science and Engineering **74** (2015) 012017 doi:10.1088/1757-899X/74/1/012017

Based on these criteria, the sizes of the areas studied in this work were 10x40 μm and 5x20 μm. In our case the maximum time of exposure for the small and the large grooves was 30 and 60 min, respectively. It should be noted that the width of the etch groove varies depending on ion current which is measured during etching.

For the experiment, a 100 μm beam with energy 30 keV was chosen. With these beam parameters, the ion current varied within 50 to 56.5 pA and was measured after each etching with a Faraday cup of the FIB column. At these current magnitudes, the beam diameter at the sample surface is approx. 25 μm which is satisfactory from the viewpoint of etching accuracy requirements.

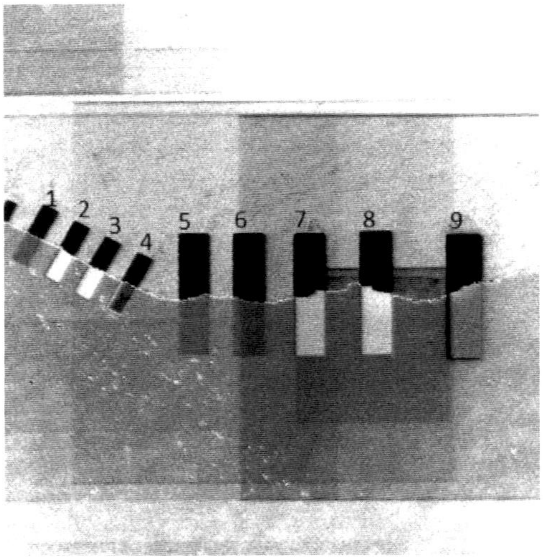

Figure 1. Secondary electron image of the sample №1 surface with etched grooves.

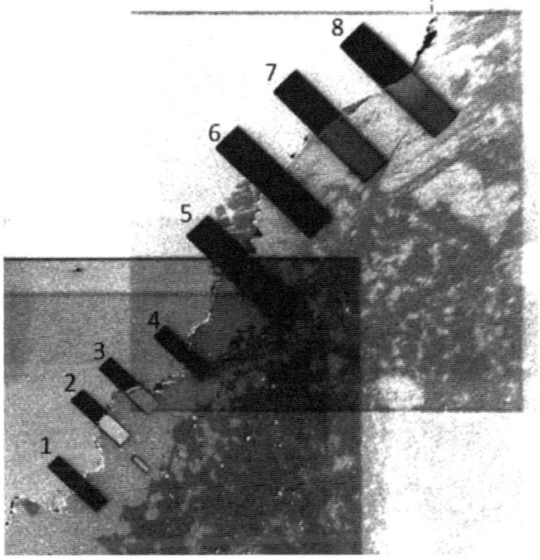

Figure 2. Secondary electron image of the sample №2 surface with etched grooves.

After ion beam irradiation, secondary ion images of the surface portions were obtained for each sample containing all the irradiated areas. These images allow assessing the sizes and positions of the etched areas and simplify detection in the scanning probe microscope. Figures 1, 2 show the images of samples Nos. 1 and 2.

Data on exposure times for etch grooves of Samples Nos. 1 and 2 are summarized in Table 1.

4. Results and Discussion

A typical image showing the topology of the ion beam irradiated areas is shown in Figure 3. For both of the samples the boundary of the diamond-like coating is well observed, and the impact area has clear limits. Some high peaks in the image are reflections of noise during probe microscope scanning.

Surface analysis of the images shows that silicon etching rate is several times that of the coating rate. For quantitative analysis of the sputtering rate, the average depth of etched areas was measured on the coating and on the silicon. The results are presented as groove depths measured using the probe method as functions of irradiation time, Figure 4.

The plot shows experimental results on etch groove depth in the silicon as a function of exposure time for samples № 1 and № 2 (curves 1 and 2, respectively). In addition, curves 3 and 4 show etch groove depth in the diamond-like coating as a function of exposure time for samples № 1 and № 2, respectively. Depths of each groove for samples № 1 and № 2 are summarized in Table 1.

The depths of the 10x40 μm grooves were close to those arrived at through calculation, and for the 5x20 μm grooves the depths were slightly greater than calculated, however, the difference was within

10%, this being probably accounted for by different ion beam movement velocities which were dependent on the shape and size of the exposed area. The sputtering rate of the DLC coating was several times lower than that of silicon.

Table 1. Exposure time and depth for each groove of samples №1 and №2.

Groove №	Milling time (sec)	Groove depth of sample №1, nm		Groove depth of sample №2, nm	
		Si	DLC	Si	DLC
		5×20 μm			
1	60	4	*[a]	*[a]	*[a]
2	300	46	1	45	*[a]
3	900	162	42	172	45
4	1800	360	150	352	156
		10×40 μm			
1	60	*[a]	*[a]	*[a]	*[a]
2	300	*[a]	*[a]	11	-5[b]
3	900	37	*[a]	35	-6[b]
4	1800	72	17	75	16
5	3600	158	40	-	-

[a] Groove depth cannot be measured using scanning probe microscopy.
[b] Foaming effect.

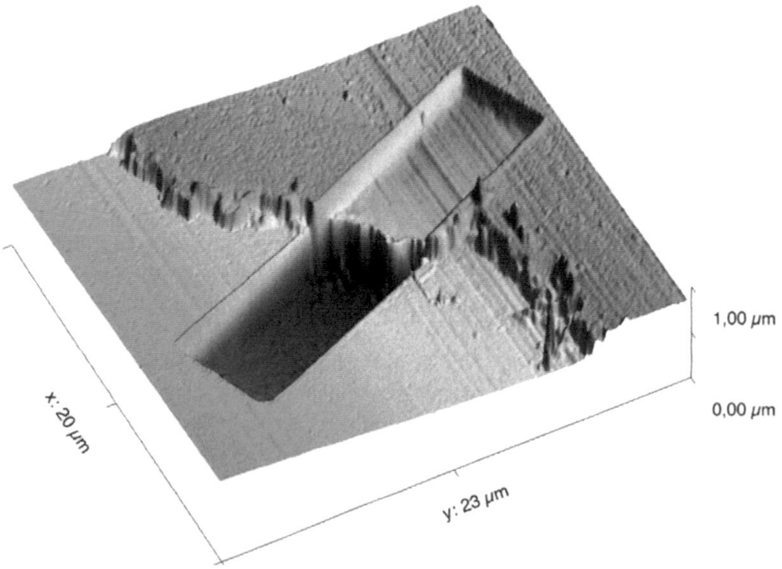

Figure 3. Model of ion irradiated area topology constructed based on the results of probe microscopy studies.

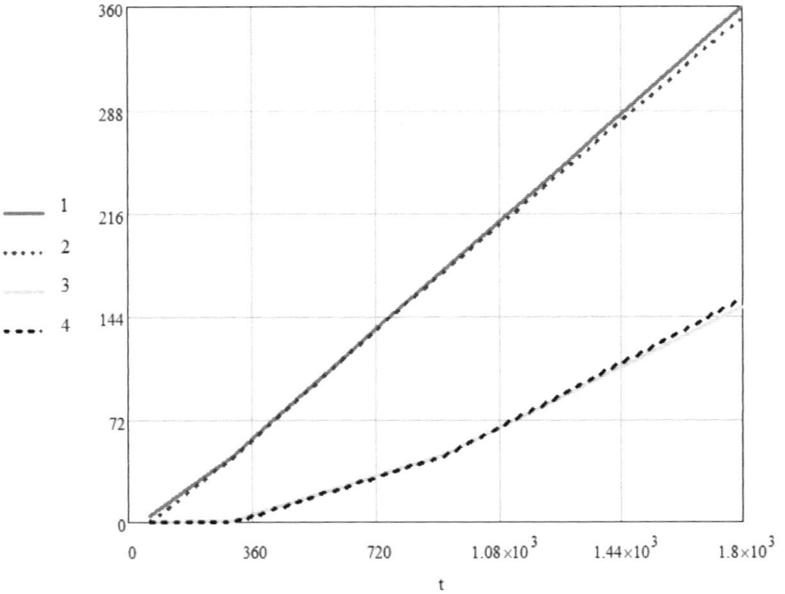

Figure 4. 5x20 um Groove Depth as a Function of Etching Time.

For both samples the dependency of silicon etch groove depths on silicon exposure time are similar and can be approximated with a straight line: that suggests a constant etching rate. Furthermore, the results proved to be almost identical for both samples, which is indicative of the stable repetitiveness of the results.

The diamond-like coating etching rate is also constant. The 'kink' in the curve corresponds to the DLC coating / Ti sublayer boundary. At an early stage of etching, both samples exhibited 'foaming' of the DLC coating which was especially clearly seen in the 10x40 μm areas and reached 6 nm for sample №2. This phenomenon has not yet been discussed in the literature and will be further studied.

Diamond-like coatings are an amorphous mixture of allotropic carbon modifications: graphite (sp2-hybridization) and carbon (sp3-hybridization). Because the density of graphite is 1.7 times lower than that of diamond, DLC coatings have quite an inhomogeneous structure from the viewpoint of density which is dependent on the degree of diamond likeness. It is therefore impossible to calculate the coating etching ratio directly, thus, the stability of the DLC coating was expressed as the ratio of the coating etching rate and the silicon etching rate which is calculated using the formula

$$SR = \frac{h_{Si}/t}{h_{DLC}/t} = \frac{h_{Si}}{h_{DLC}}|_t, \quad (1)$$

where SR is the ratio of silicon and diamond-like coating etching rates, h_{Si} and h_{DLC} are the silicon and diamond-like coating etching depths and t is etching time.

When selecting data for the calculation of SR one should take into account those etch grooves where the diamond-like coating etching depth is not greater than the coating thickness. This is required in order to exclude grooves in which both the diamond-like coating and the Ti sublayer were etched. Finally, grooves 3, 8 and 9 in sample №1 and grooves 3 and 8 in sample №2 were chosen. The value of SR was 4.1, thus, one can conclude that the diamond-like coating is approximately four times as stable against particle irradiation compared with a single crystal silicon provided that physical sputtering occurs. The results also show that the SR of the DLC coating of samples № 1 and 2 are similar.

Furthermore, the study showed that for smaller exposure areas (5x20 um compared with 10x40 um) the etching rate at depths of greater than 300 nm does not change due to back-deposition. It is therefore recommendable to choose 5x20 μm exposure areas during further studies. This will accelerate experiments due to the following reasons:

- Reduction of FIB exposure time
- Reduction of time of scanning probe microscopy studies

One should note that this work dealt only with two coating samples and was mainly aimed at testing the method and choosing optimum parameters of studies. Further studies should primarily target the dependence of etching rate on the degree of diamond likeness and other structural parameters of DLC coatings as well as the effect of coating application techniques.

5. Conclusions

A method for assessing the stability of thin coatings against particle radiations has been developed. The Ga ion irradiation stability of diamond-like coatings applied onto single crystal silicon using the vacuum electric arc method with plasma flow separation has been studied. We show that the Ga ion irradiation stability of the diamond-like coating is 4 times that of single crystal silicon.

References

[1] Mercea P, Muresan L, Mecea V and Membrane J (1985), Permeation of gases through metallized polymer membranes, *Journal of membrane science,* **24** (3) 297-307.
[2] Erlat AG, Henry BM, Ingram JJ, Mountain DB, McGuigan A and Howson RP (2001), Characterisation of aluminium oxynitride gas barrier films, *Thin solid films,* **388** (1-2) 78-86.
[3] Baek SM, Shirafuji T, Saito N and Takai O (2011), Fabrication of transparent protective diamond-like carbon films on polymer, *Jpn. J. Appl. Phys.,* **50** (8) **5** 08JD08.
[4] Abbasa GA, McLaughlina JA and Harkin-Jones E (2004), A study of ta-C, a-C:H and Si-a:C:H thin films on polymer substrates as a gas barrier, *Diamond & Related Materials,* **13** 1342-1345.
[5] Tashiro H, Nakaya M and Hotta A (2013), Enhancement of the gas barrier property of polymers by DLC coating with organosilane interlayer, *Diamond & Related Materials,* **35** 7-13.
[6] Bashkov V, Osipkov A, Mikhalev P and Makeev M (2013) Degradation behaviour and protection methods of polymeric composite materials in a Space Environment, *Proceedings of the Third International Conference on Advanced Composite Materials and Technologies for Aerospace Applications,* Glyndŵr University, Wrexham, UK, May 13 – 16, 4.
[7] Ha PCT, McKenzie DR, Bilek MMM, Kwok SCH, Chu PK and Tay BK (2007), Raman spectroscopy study of DLC films prepared by RF plasma and filtered cathodic arc *Surface & Coatings Technology* **201** 6734–6736.
[8] Srisanga C, Asanithi P, Siangchaew K, Pokaipisita A and Limsuwana P (2012), Characterization of SiC in DLC/a-Si films prepared by pulsed filtered cathodic arc using Raman spectroscopy and XPS, *Applied Surface Science* **258** 5605–5609.
[9] Yamamura Y, Tawara H (1996), "Energy dependence of ion-induced sputtering yields from monatomic solids at normal incidence." *Atomic Data and Nuclear Data Tables*, **62**(2) 149-253.
[10] Yamamura Y et al (1983), IIPJ-AM-26, Institute of Plasma Physics, Nagoya University.

AUTHOR INDEX